METAL IONS IN
BIOLOGICAL SYSTEMS

VOLUME 21

Applications of Nuclear Magnetic Resonance
to Paramagnetic Species

METAL IONS IN BIOLOGICAL SYSTEMS

Edited by

Helmut Sigel
Institute of Inorganic Chemistry
University of Basel
CH-4056 Basel, Switzerland

with the assistance of Astrid Sigel

VOLUME 21
Applications of Nuclear Magnetic Resonance
to Paramagnetic Species

MARCEL DEKKER, INC. New York and Basel

Library of Congress Catalog Number: 79-640972

MARCEL DEKKER, INC.
270 Madison Avenue, New York, New York 10016

ISSN: 0161-5149
ISBN: 0-8247-7592-9

Current printing (last digit):
10 9 8 7 6 5 4 3 2 1

PRINTED IN THE UNITED STATES OF AMERICA

Preface to the Series

Recently, the importance of metal ions to the vital functions of living organisms, hence their health and well-being, has become increasingly apparent. As a result, the long-neglected field of "bioinorganic chemistry" is now developing at a rapid pace. The research centers on the synthesis, stability, formation, structure, and reactivity of biological metal ion-containing compounds of low and high molecular weight. The metabolism and transport of metal ions and their complexes is being studied, and new models for complicated natural structures and processes are being devised and tested. The focal point of our attention is the connection between the chemistry of metal ions and their role for life.

No doubt, we are only at the brink of this process. Thus, it is with the intention of linking coordination chemistry and biochemistry in their widest sense that the series METAL IONS IN BIOLOGICAL SYSTEMS reflects the growing field of "bioinorganic chemistry." We hope, also, that this series will help to break down the barriers between the historically separate spheres of chemistry, biochemistry, biology, medicine, and physics, with the expectation that a good deal of the future outstanding discoveries will be made in the interdisciplinary areas of science.

Should this series prove a stimulus for new activities in this fascinating "field," it would well serve its purpose and would be a satisfactory result for the efforts spent by the authors.

Fall 1973   Helmut Sigel

Preface to Volume 21

In recent years, the number of metalloproteins recognized as containing paramagnetic metal ions in their active sites has increased steadily. For example, heme proteins are widely distributed in both the animal and plant kingdoms; they are key components, intimately related to cellular bioenergetics. Applications of nuclear magnetic resonance to paramagnetic species with the aim of obtaining new structural insights and identifying binding sites of metal ions are also being developed quickly. Hence, this volume is an attempt to summarize recent advances and to facilitate a wider use of NMR in studies of paramagnetic species.

In the first two chapters the theories--including the most recent developments--regarding the use of nuclear relaxation as a source for structural information are outlined and applied to examples. Next, studies of magnetically coupled metalloproteins, of paramagnetic heme proteins, and of metal-porphyrin-induced dipolar shifts for conformational analysis are summarized and critically evaluated. The sixth and final chapter is stimulated by the recent advances in NMR imaging; it considers the potential of paramagnetic ions as agents for enhancing contrast in these images: the relevance to the ultimate in biological systems, that is, to human medicine, is evident.

Finally, it should be stressed that application of NMR to paramagnetic species is only one aspect of a more general approach. There are related methods, well suited to provide further and partly complementary insights; these are covered in the following Volume 22 of this series, entitled "Endor and Electron Spin Echo for Probing Coordination Spheres."

Helmut Sigel

Contents

Chapter 6

RELAXOMETRY OF PARAMAGNETIC IONS IN TISSUE   229

Seymour H. Koenig and Rodney D. Brown III

Contributors

Numbers in parentheses indicate the pages on which the authors' contributions begin.

Ivano Bertini Department of Chemistry, University of Florence, I-50121 Florence, Italy (47)

Rodney D. Brown, III IBM Thomas J. Watson Research Center, Yorktown Heights, New York 10598 (229)

Nigel J. Clayden The Inorganic Chemistry Laboratory, University of Oxford, Oxford OX1 3QU, United Kingdom (187)

Seymour H. Koenig IBM Thomas J. Watson Research Center, Yorktown Heights, New York 10598 (229)

Claudio Luchinat Department of Chemistry, University of Florence, I-50121 Florence, Italy (47)

Michael J. Maroney Department of Chemistry, University of Massachusetts, Amherst, Massachusetts 01003 (87)

Luigi Messori Department of Chemistry, University of Florence, I-50121 Florence, Italy (47)

Geoffrey R. Moore* The Inorganic Chemistry Laboratory, University of Oxford, Oxford OX1 3QU, United Kingdom (187)

Gil Navon Chemistry Department, Tel Aviv University, Ramat-Aviv, 69978 Tel-Aviv, Israel (1)

Lawrence Que, Jr. Department of Chemistry, Kolthoff and Smith Halls, University of Minnesota, Minneapolis, Minnesota 55455 (87)

James D. Satterlee Department of Chemistry, Clark Hall 103, The University of New Mexico, Albuquerque, New Mexico 87131 (121)

---

*Present affiliation: The School of Chemical Sciences, University of East Anglia, Norwich NR4 7TJ, United Kingdom

Gianni Valensin Department of Chemistry, University of Siena,
     I-53100 Siena, Italy (1)

Glyn Williams* The Inorganic Chemistry Laboratory, University
     of Oxford, Oxford, OX1 3QR, United Kingdom (187)

---

*Present affiliation: Department of Chemistry, University College
London, 20 Gordon Street, London WC1 OAJ, United Kingdom

Contents of Other Volumes

---

*Out of print

_____

*Out of print

Other volumes are in preparation.

Comments and suggestions with regard to contents, topics, and the
like for future volumes of the series would be greatly welcome.

METAL IONS IN
BIOLOGICAL SYSTEMS

VOLUME 21

Applications of Nuclear Magnetic Resonance
to Paramagnetic Species

.

1

Nuclear Relaxation Times as a Source of Structural Information

Gil Navon
School of Chemistry
Tel-Aviv University
69978 Tel-Aviv, Israel

and

Gianni Valensin
Chemistry Department
University of Siena
I-53100 Siena, Italy

1. THEORETICAL CONSIDERATIONS

1.1. Introduction

Unpaired electrons in paramagnetic molecular species can act as the source for relaxation to magnetic nuclei. Two different interactions can occur: the through-space dipolar coupling and the through-bond contact hyperfine coupling. Both of them yield contributions to the nuclear relaxation proportional to the squares of the interaction energies. The Hamiltonian describing the energy of electron-nucleus coupling is time-dependent and its correlation function is commonly assumed to follow exponential decay with time constants $1/\tau_c$ or $1/\tau_e$ for the dipolar and scalar Hamiltonians, respectively: τ_c and τ_e have a clear physical meaning since processes like reorientational tumbling, chemical exchange, and the electron spin relaxation can be simultaneously effective in modulating the electron-nucleus coupling interactions. The following equations can therefore be given:

$$\tau_{c1}^{-1} = \tau_R^{-1} + \tau_{e1}^{-1} \qquad \tau_{e1}^{-1} = \tau_M^{-1} + \tau_{s1}^{-1}$$

$$\tau_{c2}^{-1} = \tau_R^{-1} + \tau_{e2}^{-1} \qquad \tau_{e2}^{-1} = \tau_M^{-1} + \tau_{s2}^{-1} \qquad (1)$$

Here τ_{c1}, τ_{c2}, τ_{e1}, and τ_{e2} have been introduced in order to distinguish between modulation by longitudinal or transverse electron spin relaxation (τ_{s1}, τ_{s2}). τ_R is the reorientational correlation time and τ_M is the lifetime of the electron-nucleus coupling in exchanging systems. In fact the electron-nucleus interaction energy is so big that, even if reduced by the very fast modulating motions, the nuclear relaxation rates (T_{1M}^{-1} and T_{2M}^{-1}) in the immediate neighborhood of the unpaired electrons are very fast. The consequently broadened nuclear resonances would escape detection and the paramagnetic effect would lose any structural applicability. In exchanging systems, however, the bulk "free" environment can be overwhelmingly more abundant than the bound one: chemical exchange can therefore be thought of as a carrier of information from the paramagnetic environment to the bulk. In the case of two environments (free and bound), the approach of Swift and Connick [1] and that of Luz and Meiboom [2] yield the following equations.

$$\frac{1}{T_{1obs}} = \frac{P_f}{T_{1f}} + \frac{P_b}{T_{1M} + \tau_M} \tag{2}$$

$$\frac{1}{T_{2obs}} = \frac{P_f}{T_{2f}} + \frac{P_b}{\tau_M} \left[\frac{T_{2M}^{-2} + (T_{2M}\tau_M)^{-1} + \Delta\omega_M^2}{(T_{2M}^{-1} + \tau_M^{-1})^2 + \Delta\omega_M^2} \right] \tag{3}$$

where τ_M, T_{1M}, T_{2M}, and $\Delta\omega_M$ are the exchange lifetime, the relaxation times, and the induced chemical shift of the nuclei bound to the paramagnetic species. P_f and P_b are the fractions of the nuclear species in the free and bound environments, respectively ($P_f + P_b = 1$). An elegant and simple derivation of Eqs. (2) and (3) was given by Leigh [3]. In the case where $\Delta\omega_M$ is much smaller than T_{2M}^{-1} or τ_M^{-1}, Eq. (3) takes the same form as Eq. (2). Since, usually, $P_b \ll 1$, and $P_f \cong 1$, the paramagnetic contribution to the relaxation rate is defined as follows:

$$\frac{1}{T_{ip}} = \frac{1}{T_{iobs}} - \frac{1}{T_{if}}; \qquad i = 1,2 \tag{4}$$

where T_{if} is the relaxation time of the proper control without the paramagnetic contribution. Thus, for the case of $\Delta\omega_M \ll T_{2M}^{-1}$ or τ_M^{-1},

$$\frac{1}{T_{ip}} = \frac{P_b}{T_{iM} + \tau_M}; \qquad i = 1,2 \tag{5}$$

Two possible limiting conditions can be defined. In the slow exchange limit, where $\tau_M \gg T_{iM}$, $T_{ip}^{-1} = P_b\tau_M^{-1}$, the relaxation rate giving only kinetic information. In the fast exchange limit $T_{iM} \gg \tau_M$. Then $T_{ip} = P_b T_{iM}^{-1}$ and the information obtained concerns T_{iM}, which is related to structural parameters. In the case of three or more environments the above approach can easily be extended, as shown in [4,5].

1.2. Relaxation Theory

The theory for paramagnetic relaxation reported here and available
in many books and review articles [6-10] applies to all the nuclei as
well as to all the species containing unpaired electrons. However,
it will be shown hereafter that different information is obtained
from different nuclei within the same molecule and that suitable
refinements have to be adopted when dealing with interaction with
specific paramagnetic metal ions or free radicals.

The relaxation theory originally developed by Solomon and
Bloembergen, denoted as SB hereafter [11-13], under a point-dipole
approximation and by assuming the existence of only one electronic
relaxation time was extended for one τ_{s1} and one τ_{s2} [14-16]. The
following equations were derived:

$$T_{2M}^{-1} = \frac{2}{15}\gamma_I^2\gamma_S^2\hbar^2 S(S+1)r^{-6}\left\{\frac{3\tau_{c1}}{1+\omega_I^2\tau_{c1}^2} + \frac{7\tau_{c2}}{1+\omega_s^2\tau_{c2}^2}\right\}$$

$$+ \frac{2}{3}A^2 S(S+1)\left\{\frac{\tau_{e2}}{1+\omega_s^2\tau_{e2}^2}\right\} \tag{6}$$

$$T_{2M}^{-1} = \frac{1}{15}\gamma_I^2\gamma_S^2\hbar^2 S(S+1)r^6\left\{4\tau_{c1} + \frac{3\tau_{c1}}{1+\omega_I^2\tau_{c1}^2} + \frac{13\tau_{c2}}{1+\omega_s^2\tau_{c2}^2}\right\}$$

$$+ \frac{1}{3}A^2 S(S+1)\left\{\tau_{e1} + \frac{\tau_{e2}}{1+\omega_s^2\tau_{e2}^2}\right\} \tag{7}$$

where γ_I and γ_S are the nuclear and electronic gyromagnetic factors,
S is the electronic spin quantum number, r is the electron nucleus
separation, and A is the hyperfine interaction constant. The first
term in parentheses in both equations is the explicit form of the
electron-nucleus dipolar interaction which provides the desired
structural information. The second term depends on the electron
delocalization onto the nuclei. It is less easily interpreted, but
it is negligible in many cases of interest.

In addition, dipolar interactions with the paramagnetic species modulated by free translational diffusion and the electron spin relaxation must be also considered (outer sphere relaxation). This mechanism is expected to yield the most relevant relaxation pathway for nuclei excluded from the immediate neighborhood of the unpaired electrons. The contributions of such a mechanism to T_1 were given by Pfeifer [17,18] while an extension to T_2 was also formulated [19]. These are given in Eqs. (8) and (9).

$$T_{1os}^{-1} = \gamma_I^2\gamma_S^2\hbar^2 S(S+1)\left\{\frac{3}{2}I(\omega_I,\tau_{s1}) + \frac{7}{2}I(\omega_s,\tau_{s2})\right\} \qquad (8)$$

$$T_{2os}^{-1} = \gamma_I^2\gamma_S^2\hbar^2 S(S+1)\left\{I(0,\tau_{s1}) + \frac{3}{4}I(\omega_I,\tau_{s1}) + \frac{12}{4}I(\omega_s,\tau_{s2})\right\} \quad (9)$$

In Eqs. (8) and (9), the power spectrum function $I(\omega,\tau)$ depends also on the translational correlation time, and is proportional to C/r_{os} where r_{os} is the minimal distance between the nuclei of the bulk ligands and the paramagnetic center, averaged over all the possible mutual orientations, and C is the concentration of the paramagnetic ions. The explicit form of $I(\omega,\tau)$ was given by Pfeifer [17].

The modes of working out the relaxation theory or its later refinements will be exemplified hereafter, but it is now of concern to note that different nuclei of the same nuclear species or of different nuclear species within a given molecule can experience quite different electron-coupling energies due to variation of $1/r^6$, A, τ_R, and γ_I. It is therefore a common case that both slow and fast exchange conditions are obeyed within a given paramagnetic adduct, yielding simultaneous information about kinetic and structural parameters. In the same way different paramagnetic centers may also provide very different interaction energies, even in adducts of similar structure, due to variations of S, A, τ_{s1}, and τ_{s2}. It is well known indeed that paramagnetic metal ions having relatively long electronic relaxation times, such as Mn^{2+}, Fe^{3+}, Cr^{3+}, and Gd^{3+}, as well as organic free radicals constitute the class of relaxation probes, as opposed to that of shift probes. As a matter of fact, only relaxation probes will be of concern herein and the above reported exchange

conditions assume that $\Delta\omega_M$ (the shift between free diamagnetic and
bound paramagnetic sites) is much smaller than either τ_M^{-1} or T_{2M}^{-1} .

The first refinement needed by the relaxation theory of Solomon
and Bloembergen is related to taking into account the dependence of
the electronic Zeeman interaction on the orientation of the molecule
in respect to the external field. Such dependence is brought about
by the anisotropy of the g tensor or the hyperfine interaction tensor.
Moreover, the principal axes of the g and A tensors may not coincide
and the A tensor need not be symmetric or diagonal. In every case it
ought also to be distinguished between fast or slow rotational rates
as compared with the electron spin relaxation rates, yielding time-
averaged or space-averaged relaxation times. Although early refinements or new derivations were carried out [20-22], analytical expressions for the electron and nuclear relaxation times in liquid solution
were recently derived for the general case of an anisotropic g tensor
with sets of principal axes not coincident with those of an hyperfine
interaction tensor A, having the nine components all different [23].
The derived expressions were used to interpret the nuclear relaxation
times in the Ru(acac)$_3$ complex. However, it turned out that problems
aiming from either g or A anisotropies affect the nuclear relaxation
rates only to a very small extent, even in the limit of slow motion
[24].

It should be mentioned [25] that for d^5 and f^7 ions which are
S-state ions g is expected to be isotropic. The anisotropicities in
the electronic transitions are derived from the zero field splitting
(ZFS), which is discussed below. The anisotropies of g and A may be
important for ^{13}C, ^{15}N, and ^{17}O nuclei since the unpaired electron
can reside in p orbitals. For ^{1}H and ^{2}H, which are S-state atoms,
the hyperfine interaction is expected to be mostly isotropic. It
should be mentioned, however, that some anisotropy is obtained from
spin densities on p orbitals of neighboring oxygen or nitrogen atoms
contributing to pseudocontact shifts [26].

The application of the relaxation theory to NMR relaxation
studies has, on the other hand, encountered several problems arising

from the associated expressions for the electron spin relaxation, as
formulated by Bloembergen and Morgan [27]:

$$\frac{1}{\tau_{s1}} = B\left\{\frac{\tau_v}{1 + \omega_s^2\tau_v^2} + \frac{4\tau_v}{1 + 4\omega_s^2\tau_v^2}\right\} \tag{10}$$

where τ_v is the correlation time responsible for modulation of transient or static zero field splitting and B is a constant related to
the zero field splitting parameters. From the electronic Zeeman
energy levels of $S > 1/2$ ions it is obvious that the electronic spin
relaxation is governed by more than one electronic relaxation time.
For $S = 1$ and $S = 3/2$ ions there are two T_1 and two T_2 electronic
relaxation times. For $S = 5/2$ there are three electronic T_1's and
three T_2's. Each one of these electronic relaxation times has a
different contribution to the average electronic relaxation time
which affects the nuclear relaxation in the Solomon-Bloembergen
equations. A detailed derivation of these contributions using Redfield's theory was given by Rubinstein et al. [28]. Their numerical
results for the transverse and longitudinal electronic relaxation
times, as well as their corresponding weights, as a function of $\omega\tau$,
are reproduced in Figures 1 and 2. It is interesting to note that
within the limit of this theory the Bloembergen and Morgan's expression does represent the average longitudinal electronic relaxation
time with $B = 32/25\Delta^2$, where $\Delta^2 = 3/2D^2 + 2E^2$ is the sum of the
diagonal elements of the ZFS.

An expression for T_{2s} analogous to the Bloembergen and Morgan
Eq. (10) may be written as follows [19,29]:

$$\frac{1}{\tau_{s2}} = \frac{B\tau_v}{2}\left\{3 + \frac{5}{1 + \omega_s^2\tau_v^2} + \frac{2}{1 + 4\omega_s^2\tau_v^2}\right\} \tag{11}$$

In fact, a comparison of this equation with the numerical result of
Rubinstein et al. [28] gave a good agreement [19]. τ_{s2} may also be
estimated from the EPR linewidths. However, this procedure should
be done at the same external magnetic field, and one must ascertain
that inhomogeneous broadening does not affect the EPR linewidth.
Using the Solomon-Bloembergen-Morgan theory as given in its original

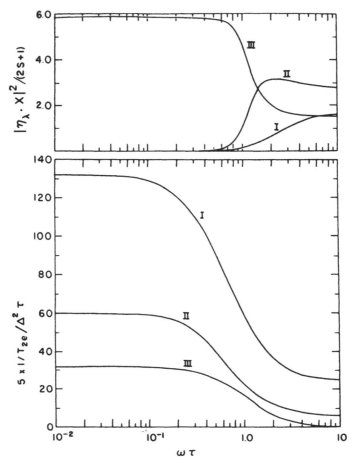

FIG. 1. Plots of the eigenvalues $1/T_{2e}$ and the corresponding
intensities for S = 5/2. Reproduced by permission from [28].

formulation thus provides only a very simplified approach to systems
with S > 1/2, where well-defined τ_{s1}'s and τ_{s2}'s do not exist. This
occurs when the Redfield limit ($\tau_v \ll \tau_{s2}$) is no longer obeyed by the
electron spin system ("slow-motion problem"). Such a situation must
be faced when the time modulation of the zero field splitting is equal
to or longer than the electron relaxation times or when the zero field
splitting is of the same order of magnitude as the electron Zeeman
energy. The slow-motion problem has been dealt with in several experimental NMR studies [29-34], where the applicability of the Solomon-

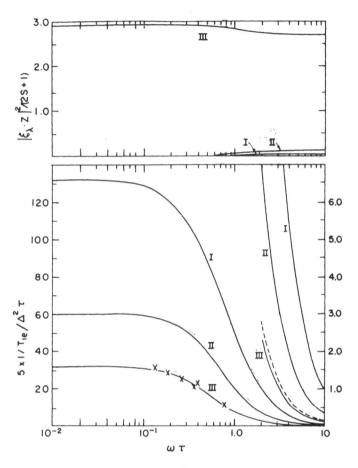

FIG. 2. Plots of the eigenvalues $1/T_{1e}$ and the corresponding intensities for S = 5/2. The crosses are best fits for the Mn^{2+} electronic relaxation times obtained from proton relaxation data. For large $\omega\tau$, plots are also given on an enlarged scale (given on the right coordinate). The dashed curve in this region is for the average $\langle 1/T_{1e}\rangle$. (Reproduced by permission from [28].)

Bloembergen-Morgan theory was thoroughly challenged and refinements or new derivations were suggested. Recently, a new formulation of the theory has appeared [9,35,36] which circumvents the concepts of electron spin relaxation and is valid in the slow-motion regime for the electron spin. Westlund et al. [36] developed a theory for nuclear spin-lattice and spin-spin relaxation for $S \geqslant 1$ electronic spin, taking into account the effects of the zero field splitting (ZFS) and extending the theory to the slow-motion region. The behavior of both T_1 and T_2 were found to deviate significantly from that predicted by Solomon and Bloembergen equations at the region of ZFS larger than the Zeeman interaction. It is interesting to note that for high field, where the Zeeman interaction is greater than the ZFS, the Solomon and Bloembergen equation for T_1 applies even in the region of slow motion.

In this new theory the electron spin is considered to be a part of the lattice, whereas the spin systems contain only the nuclear spin; while the description of the lattice becomes more complicated, all the fast motions are included in the lattice and the electron spin relaxation times do not need to be specified. In their treatment they assumed that the only contribution to the electron spin relaxation is due to a zero field splitting inter-action modulated by the molecular rotational diffusion motion. The nuclear relaxation rates are expressed in terms of the complex spectral densities,

$$T_{I1}^{-1} = -2R_e k_{1,-1}(-\omega_I) \tag{12}$$

$$T_{I2}^{-1} = R_e \left\{ k_{o,0}(0) - k_{1,-1}(-\omega_I) \right\} \tag{13}$$

which are Fourier-Laplace transforms of the appropriate correlation functions. The spectral densities may be expressed as sums of three terms corresponding to the dipolar interaction (K^{DD}), the scalar interaction (K^{SC}), and the cross-term between the two interactions (K^{DD-SC}). Spectral density charts of K^{DD}, K^{SC}, and K^{DD-SC} were computed, as shown in Figures 3 and 4 for $S = 3/2$ and $S = 5/2$. The analysis of the K^{DD} chart (Fig. 3) provided evidence that, in the region where the Zeeman interaction (ω_s) is large compared with the ZFS interaction (f_o), the Solomon approach is valid even outside the narrowing condition if the Zeeman interaction is large enough. On the contrary, the Solomon equation breaks down when the ZFS inter-action is larger than the Zeeman interaction. The cross-term spec-tral density chart (Fig. 4) shows that K^{DD-SC} vanishes in the Redfield

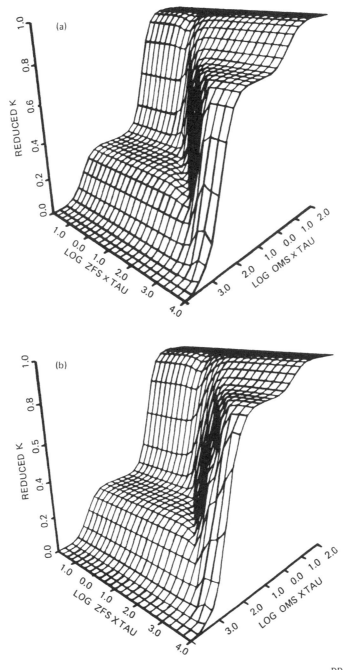

FIG. 3. Plots of the completely reduced spectral density $\tilde{K}_{1,-1}^{DD}$ for (a) S = 3/2 and (b) S = 5/2. (Reproduced by permission from [36].)

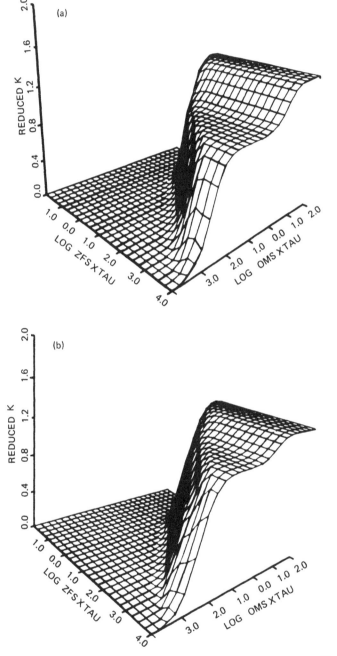

FIG. 4. Plots of the completely reduced spectral density $\tilde{k}_{1,-1}^{DD-SC}$ for
(a) S = 3/2 and (b) S = 5/2. (Reproduced by permission from [36].

region for the electron spin. The diagonal $f_0 = \omega_S$ therefore acts as a dividing line between the region where the SB approach is valid and the slow-motion region. It was shown [37] that the new theoretical model yields a consistent set of reasonable values for the ^{15}N relaxation rates in the complex $Ni(dpm)_2(aniline)_2$ which could not be fitted on the basis of the modified Solomon-Bloembergen equations.

1.3. Curie Spin Relaxation

The nuclear relaxation described by the Solomon and Bloembergen equations is caused by the time dependence of the dipolar and contact magnetic fields of the electron spin acting on the nucleus. In addition to the exchange lifetime and, for the dipolar interaction, also the molecular rotations, these magnetic fields are modulated by the relaxation of the electronic spin. For macromolecules in solutions, at high magnetic fields another relaxation mechanism should be considered, i.e., the dipolar field of the thermally averaged electronic spin—the "Curie spin" [38,39]. It is aligned with the external magnetic field and its value is:

$$\langle S_c \rangle = \frac{g\beta S(S + 1)B_0}{3kT} \qquad (14)$$

Its dipolar interaction with the nuclei is modulated by the molecular rotation τ_R and by the chemical exchange, i.e., by τ_D given by:

$$\tau_D^{-1} = \tau_R^{-1} + \tau_M^{-1} \qquad (15)$$

Since this relaxation mechanism is not dependent on the electronic spin relaxation time, its contribution becomes apparent in systems where τ_S is relatively short. Thus, the modified Solomon-Bloembergen equations, after dropping the terms involving the contact interaction and those involving ω_S and τ_{C2}, which are usually negligible at high fields, are:

$$T_{1M}^{-1} = \frac{6}{5}\gamma_I^2\gamma_S^2\hbar^2 S(S + 1)r^{-6}\left\{S_c^2 \frac{\tau_D}{1 + \omega_I^2\tau_D^2} - [\frac{1}{3}S(S + 1) - S_C^2]\frac{\tau_{C1}}{1 + \omega_I^2\tau_{C1}^2}\right\}$$

$$(16)$$

$$T_{2M}^{-1} = \frac{1}{5} \gamma_I^2 \gamma_S^2 \hbar^2 S(S+1) r^{-6} \left\{ S_C^2 (4\tau_D + \frac{3\tau_D}{1 + \omega_I^2 \tau_D^2}) \right.$$

$$\left. + [\frac{1}{3} S(S+1) - S_C^2] (4\tau_{C1} + \frac{3\tau_{C1}}{1 + \omega_I^2 \tau_{C1}^2}) \right\} \qquad (17)$$

It was shown [38] that in typical cases the influence of the Curie spin on T_1 is quite small, whereas it may be dominant on the transverse relaxation.

An interesting application of the effect of the Curie spin relaxation on the proton NMR linewidths was given for high-spin iron(II) heme proteins such as deoxymyoglobin, deoxyhemoglobin, and cytochrome C'. The quadratic field dependence that was observed for all these cases [40-42] was interpreted by Livingston et al. [42] as contribution from Curie spin relaxation. From the magnetic field dependence of the linewidths they were able to derive rotational correlation times τ_R for these proteins in vitro as well as in situ in bovine heart muscle. The intracellular τ_R was estimated to be roughly 2.5 times longer than in dilute solutions.

1.4. Effects of Covalency

Extensive investigation has been dedicated to correct the relaxation theory in the several cases where the point dipole approximation for the electron spin fails and the spatial distribution of the unpaired electrons within either the metal-centered or the ligand-centered orbitals cannot be neglected. The problem was first studied by Waysbort and Navon [43] in the highly symmetric octahedral $Ru(NH_3)_6^{3+}$ complex. In this complex the metal-proton distance is accurately known and the distance obtained from the NMR relaxation data was smaller by 8%. The spatial distribution of the unpaired electron was included in Solomon's derivation of the dipolar relaxation. As a result, an equation similar to that in the point dipole approximation was obtained except that the interatomic distance was replaced by the effective distance r_{eff}, so that $r_{eff}^{-6} = <r^{-3}>^2$. Numerical calculations for the proton relaxation rates showed that the effect of the distribution of the spin density on the ruthenium d orbitals was small, whereas spin delocalization to the ligand orbitals caused a reduction of 5-12% in the effective metal-proton distance, obtained by the point dipole approach, corresponding to a factor of 1.4-2.2 in the dipolar interaction. When considering the water protons in hexa-aquo ions [44,45], the increase of the electron-nuclear dipolar inter-action by covalency was found to be relatively small owing to cancel-lations in the contributions from spin densities in the various water molecular orbitals.

Gottlieb et al. [46] repeated the same calculation using a density matrix formalism arriving at the same final results. They emphasized that dramatic deviations from the point dipole approxima-tion may occur in paramagnetic transition metal complexes of lower symmetry, where extensive spin delocalization is possible, especially for nuclei other than protons. As an example, the relaxation theory predicts T_1^H/T_1^C ratios of 0.34 and 0.18 for the beta and para positions in the complex bis(2-N-phenyl-4-N-phenylimino-1-pentene)nickel(II), whereas the experimental values are 2.4 and 1.8. Introducing ligand-centered contributions allowed them to calculate values of 3.1 and 2.2, respectively.

By following the same spin density matrix formalism, but per-forming ab initio Hartree-Fock calculations for several hexaaquo metal complexes, Nordenskiöld et al. [47] showed that the effective distance between the ^{17}O nucleus and the unpaired electron is signifi-cantly shorter than the internuclear oxygen-metal distance. Moreover, it came out that the deviation from the point dipole approximation for the ^{17}O nucleus-metal ion distance in $M(H_2O)_6^{2+}$ was a function of the metal, the deviation in the effective metal-oxygen distance being only 1% for Mn^{2+} at one end of the scale but 14-16% for Cu^{2+} and Ni^{2+} at the other end. However, the point dipole approximation was found to work very satisfactorily for the hydrogen atoms in hexaaquo ions, with deviations of the effective metal hydrogen distances ranging

from 0% for Mn^{2+} to 1% for Co^{2+}, Ni^{2+}, and Cu^{2+} in agreement with
the previous result [45]. Extension of the theoretical approach to
bismalonato-nickel(II) complexed with either water or ammonia provided evidence that appreciable deviations from the point dipole
approximation were to be found only for the ligand nuclei directly
bonded to the metal and for the carbon atoms next to the oxygen in
the malonate ion [48].

1.5. Effects of Spin Diffusion

Mutual flips of dipolar-connected nuclear spins through zero-quantum
transitions (cross-relaxation effects), when occurring at a fast rate
as compared with the spin-lattice relaxation rate, provide a very
efficient way of magnetization transfer between the coupled spins
(spin diffusion) before any energy transfer to the lattice occurs.
Since the cross-relaxation rate increases linearly with τ_c, spin
diffusion plays a dominant role in determining T_1 values for relatively large proteins, even at relatively low frequencies such as
the nuclear relaxation rate, as first pointed out by Andree [49],
Sykes et al. [50] extensively investigated the role of cross-relaxation
in ^1H T_1 measurements in proteins. The results indicated that spin
diffusion plays a dominant role in determining T_1 measurements for
relatively large proteins, even at relatively low frequencies such as
100 MHz. Theoretical treatment of spin diffusion effects in nuclear
relaxation rates of paramagnetic species has been recently reported
by Granot [51]. The recovery of the longitudinal magnetizations of
two weakly coupled nuclear spins I and J, which are part of a ligand
molecule exchanging between a diamagnetic free and a paramagnetic
bound environment, was given in the matrix form:

$$\frac{d}{dt}M = A(M - M_0) \tag{18}$$

where

$$
A = \begin{bmatrix} -\rho_{IB} & \sigma_{IJ} & \kappa_F & 0 \\ \sigma_{IJ} & -\rho_{JB} & 0 & \kappa_F \\ \kappa_B & 0 & -\rho_{IF} & \sigma_{IJ} \\ 0 & \kappa_B & \sigma_{IJ} & -\rho_{JF} \end{bmatrix} \qquad M = \begin{bmatrix} M_{IB} \\ M_{JB} \\ M_{IF} \\ M_{JF} \end{bmatrix} \qquad M_0 = \begin{bmatrix} M_{OB} \\ M_{OB} \\ M_{OF} \\ M_{OF} \end{bmatrix} \qquad (19)
$$

and where $\rho_{IB} = R_{1IF} + R_{1TP} + \sigma_{IJ} + \kappa_B$; $\rho_{IF} = R_{1IF} + \sigma_{IJ} + \kappa_F$ and analogous relations for the spin J. R_{1IF} (or R_{1JF}) is the direct relaxation rate of spin I (or J) in the diamagnetic environment, which contains ρ_{IJ} (from dipolar IJ interaction). R_{1IF}^{other} (R_{1F}^{other}) [from mechanisms other than the dipolar R_{1IP} (R_{1JP})] is the relaxation rate in the bound environment, to be equaled to T_{1M}^{-1} from the Solomon-Bloembergen theory. σ_{IJ} is the cross-relaxation rate given by [14]:

$$
\sigma_{IJ} = \frac{2}{15} \frac{\gamma_I^4 \hbar^2 I(I+1)}{r_{IJ}^6} \left\{ J_0(\omega_I - \omega_J) - 6J_2(\omega_I + \omega_J) \right\} \qquad (20)
$$

κ_F and κ_B are the first-order exchange rates from the free and bound states, respectively.

Derivation of approximate solutions for a number of limiting cases (in terms of the relative rates of exchange, cross-relaxation, and spin-lattice relaxation) showed that reliable and accurate paramagnetic relaxation rate measurements are to be expected only when the spin diffusion process is very slow in comparison with the other rates. If this is not the case, as usually found for macromolecular systems at high frequency, the recovery of longitudinal magnetization can rarely be fitted with a single exponential decay. In such cases measurements of the initial rate constant must be performed, which is better accomplished upon selective irradiation of the observed resonance.

An experimental evidence of cross-relaxation effects was given by Sletten et al. [52] for the recovery of longitudinal magnetizations

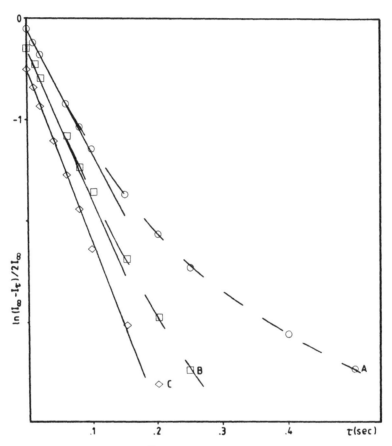

FIG. 5. The effect of Gd^{3+} on the plot of $\ln(I_\infty - I_\tau)/2I_\infty$ vs. τ for a hyperfine shifted proton resonance of the 360-MHz ^1H NMR spectrum of 5 mM horse heart ferricytochrome c in D_2O at 25°C and pH = 6.0. A, pure protein, $[Gd^{3+}] = 0$; B, $[Gd^{3+}] = 10$ mM; C, $[Gd^{3+}] = 20$ mM. The straight lines indicate the initial slopes; dashed curves indicate connectivity and have no theoretical significance. All curves are plotted to the same scale but are displaced vertically for clarity. (Reproduced by permission from [52].)

of hyperfine shifted protons in horse heart ferricytochrome c upon addition of Gd^{3+} ions as shown in Figure 5 for one resonance at different Gd^{3+} concentrations.

2. APPLICATIONS

2.1. Determination of Hydration Numbers

Since the early discovery by Eisinger et al. [53,54] that upon
binding of paramagnetic metal ions to macromolecules the water
proton relaxation rates were "enhanced," a large number of investigations have been reported dealing with metalloproteins, especially
metalloenzymes [7,55]. In several cases the macromolecule was made
paramagnetic by substituting the natural diamagnetic metal ion (e.g.,
Zn^{2+}, Mg^{2+}, Ca^{2+}, etc.) with a "paramagnetic relaxation probe" (in
most cases Mn^{2+}). The relaxation enhancement of the water proton
relaxation rates, which occurs when paramagnetic metal ions bind to
macromolecules, is specific to paramagnetic metal ions with long τ_S,
so that the predominant correlation time of the electron-nuclear
dipolar interaction is the rotational correlation time τ_R, which
becomes longer upon binding to the macromolecules.

The problem of evaluating quantitatively the parameters which
govern the relaxation was approached in different ways due to the
difficulty of accounting for all the variables in the theoretical
equations for $1/T_{1p}$ and $1/T_{2p}$. The most commonly adopted procedure
consisted of fitting the experimental results at various frequencies
to the Solomon-Bloembergen-Morgan equations. The problems arising
in the fitting procedures and the refinements of the theory are dealt
with in other chapters of this book and will therefore be neglected
herein. A different approach was based on the comparison of T_1 and
T_2 at a single, suitably chosen, frequency [56]. Under fast exchange
conditions and with the assumption that the ratio of contact and
dipolar interactions is the same as for the free ion, the ratio of
T_{1p} and T_{2p} for bound Mn^{2+} is:

$$\frac{T_{1p}}{T_{2p}} = 0.5 + \left[0.67 + 0.022\,\frac{\tau_e}{\tau_c}\right]\left[1 + \omega_I^2\tau_c^2\right] \tag{21}$$

Substituting τ_c from Eq. (21) in the equation for $1/T_{1p}$ obtained
from Eqs. (5) and (6), and using the free ion value for r, the final

expression for the hydration number q was obtained:

$$q = 3.26 \times 10^{-14} \frac{(T_{1p}/T_{2p} - 0.5)}{(T_{1p}/T_{2p} - 1.19)^{1/2}} \frac{\omega_I}{NT_{1p}} \quad (22)$$

In spite of its simplicity this approach is an oversimplification owing to the assumption of the fast exchange limit and the neglect of the outer sphere contribution, which becomes important at high frequencies [57].

An attempt for a quantitative evaluation of the structural and kinetic parameters was based on comparison of proton and deuteron spin-lattice relaxation rates in solutions of paramagnetic macromolecules in H_2O and D_2O [58]. The determination of both rates at two frequencies yields four equations which can be simultaneously solved for the four unknowns q, τ_M, $\tau_c(\omega_{I1})$, and $\tau_c(\omega_{I2})$. This method neglects any outer sphere contribution and depends on the detailed frequency dependence given by Solomon-Bloembergen-Morgan equations which, as discussed above, do not apply rigorously to macromolecular paramagnetic systems.

A combination of the above two approaches was recently suggested by Kushnir and Navon [57]. In the new method T_1 and T_2 for protons and deuterons are measured in the same solution and at one high magnetic field. A proper blank solution has to be measured in order to obtain accurate values of T_{1p} and T_{2p}, especially for the deuterons which have a significant quadrupolar relaxation in solutions of macromolecules. It was suggested to take as a blank identical solutions with the paramagnetic ion replaced by a diamagnetic metal ion such as Zn^{2+}. Having the values of four relaxation times at one magnetic field makes it possible to calculate the hydration number, exchange lifetime, and correlation time as well as the contribution from outer sphere water molecules. This method has the advantage of not relying on magnetic field dependence since there is an uncertainty as to how the electronic correlation time varies with frequency. Also the relaxation is better described by the Solomon-Bloembergen equations at high fields where the Zeeman energy is greater than the zero field splitting [9,36]. Thus, this approach is relatively free from arbitrary assumptions and indeed results in parameters, such as hydration numbers and exchange lifetimes, that are invariant at several magnetic fields. Since in this method the hydration number q is obtained in the form of q/r^6, an absolute value of q depends on the knowledge of r^{-6}. In their work Kushnir and Navon [57] used the value of q/r^6 for the free metal ion; thus, by assuming the same r for both systems, they avoided the need for prior knowledge of r and the possible effects of covalency on its determination. This method is of course strictly applicable only in cases where all water molecules of hydration have about the same exchange lifetimes and distances. Otherwise the number of unknown parameters becomes too large.

2.2. Relaxation of Bound Ligands

Perhaps the most exciting application of NMR relaxation studies of paramagnetic species is the possibility of obtaining three-dimensional structures of enzyme-substrate complexes. Such structures can only rarely be determined by x-ray crystallography. NMR is probably the only technique whereby detailed structural information can be obtained in solution. The source of such information is the NMR relaxation caused by dipolar interaction as described by Solomon-Bloembergen equations. Once the correlation times governing the dipolar interaction are known, evaluation of T_{1M} or T_{2M}—the relaxation times of the bound substrates—yields the distance r between the various nuclei and the paramagnetic center [59,60]. The hyperfine interaction in Eqs. (6) and (7) may be neglected in most cases, where the nuclei are not directly attached to the paramagnetic metal ion. The values of T_{1M} and T_{2M} are usually obtained by measuring the effect of binding of substrates on the relaxation times of the excess free substrate through chemical exchange, using Eqs. (2) and (3). In order to find out whether the relaxation is exchange-limited, the temperature dependencies of the relaxation times have been used [59-61]. However,

one should take this criterion with caution. It is generally true
that τ_M decreases with temperature, and therefore relaxation times
which are found to increase with temperature cannot be exchange-
limited. Yet the converse is not always correct: relaxation times
which decrease with temperature may be exchange-limited but may also
be in the limit of fast exchange with, for example, a case of
$\omega_I \tau_c > 1$. In general one must always consider the possibility of
intermediate exchange. A way to obtain interatomic distances in a
system with intermediate exchange without resorting to extrapolations
of temperature dependencies was suggested by Lanir and Navon [62]:
when a substrate contains several proton groups which share the same
τ_M and τ_c and differ in their distances to the paramagnetic center,
then for each proton:

$$\frac{T_{1p}}{T_{2p}} = \frac{T_{1M} + \tau_M}{T_{2M} + \tau_M} \tag{23}$$

This equation may be rearranged to:

$$\left(\frac{T_{1M}}{T_{2M}}\right) [(T_{2M} + \tau_M) - \tau_M] + \tau_M = \frac{T_{1p}}{T_{2p}}(T_{2M} + \tau_M) \tag{24}$$

From Solomon-Bloembergen equations without the contact term:

$$\frac{T_{1M}}{T_{2M}} = \frac{7}{6} + \frac{2}{3} \omega_I^2 \tau_c^2 \tag{25}$$

In Eq. (24) both $T_{2M} + \tau_M = P_B \cdot T_{2p}$ and T_{1p}/T_{2p} are experimentally
determined. Thus, it is enough to measure them for two proton groups
in the same molecule in order to obtain τ_M and T_{1M}/T_{2M} which gives τ_c
using Eq. (25). When more than two groups are present in the sub-
strate, these two constants are overdetermined. An example of a
structure of the inhibitor, sulfacetamide, bound to carbonic anhydrase,
is given in Figure 6.

A large number of NMR studies on enzyme-substrate interactions
has been done by Mildvan and his coworkers, mainly on ATP-utilizing
enzymes (for recent reviews see [63,64]). Very interesting applications have been done in their works with the substitution-inert

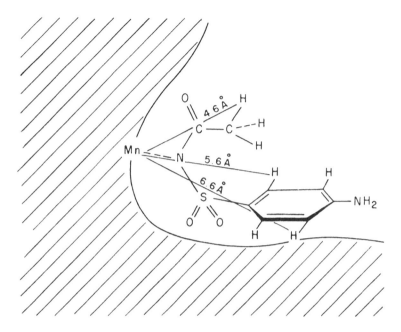

FIG. 6. A model of the carbonic anhydrase-sulfacetamide complex. (Reproduced by permission from [62].)

Co(III) and Cr(III) complexes of ADP and ATP [65-67]. It was found that the $Co(NH_3)_4ATP$ complex binds to various ATP-utilizing enzymes in much the same way as ATP, but without hydrolysis, so that ^1H and ^{31}P studies of the complex are feasible. Examples of such studies were given for cAMP-dependent protein kinase, PP-Rib-P synthetase, and other enzyme systems [68-71]. The stable Cr(III)ADP and Cr(III)ATP have the additional advantage that the Cr(III) center can serve as a relaxation agent [72-74]. A combination of proton, ^{13}C, ^{19}F, ^{31}P NMR, as well as the NMR of ^{205}Tl, which serves as a model for the monovalent cation bound to pyruvate kinase [75], with Mn^{2+}, Cr(III)ATP, and Cr(III)ADP as paramagnetic centers have led to a detailed model of the active site of pyruvate kinase [63]. Such a model and its implication for the mechanism of action of pyruvate kinase are shown in Figures 7 and 8. These results show the power of the method in describing structures of enzyme-substrate complexes

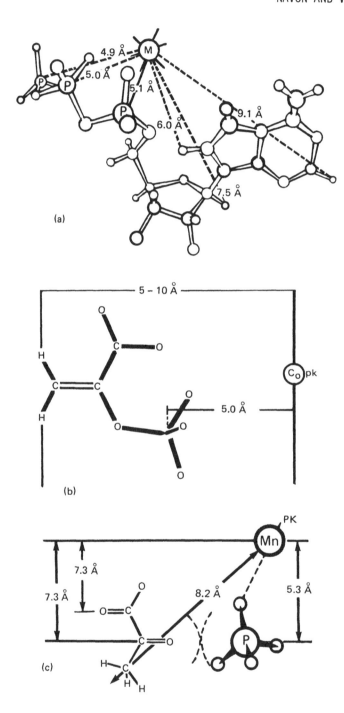

(a)

(b)

(c)

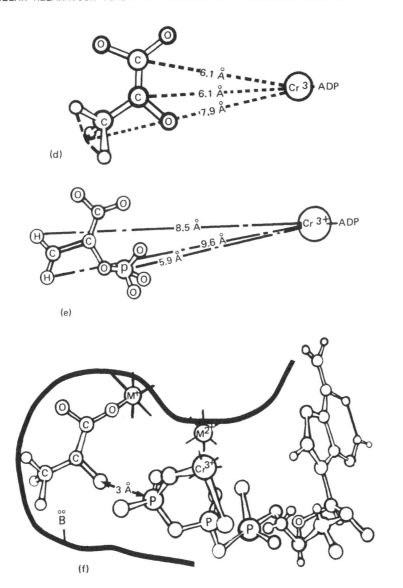

FIG. 7. The steps in construction of a model of enzyme-substrate complexes of pyruvate kinase [63]. The substrates are (a) ATP, (b) phosphoenolpyruvate, (c) pyruvate and Pi, (d) Cr(III)ATP and pyruvate, (e) Cr(III)ADP and phosphoenolpyruvate, and (f) composite of all. (Reproduced by permission from [63].)

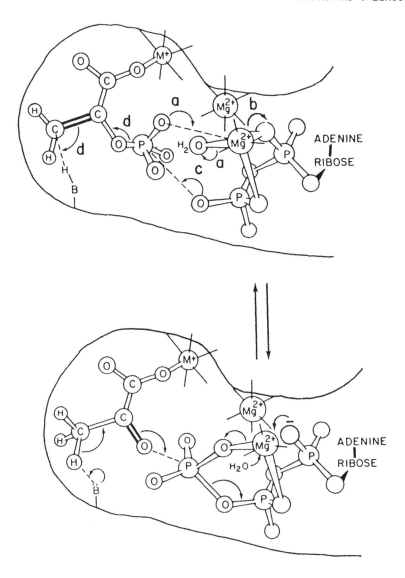

FIG. 8. Minimal mechanism of pyruvate kinase based on structural
and kinetic data [76]. The phosphoryl transfer from phosphoenol
pyruvate to MgADP occurs in four steps: (a) a water on MgADP is
replaced by the phosphoryl group of P-enolpyruvate, (b) Mg dissociates from the α-P of ADP, (c) the phosphoryl group is transferred,
(d) the enolate of pyruvate is protonated. (Reproduction by
permission from [76].)

and their mechanism of action, even for multicomponent systems.
Although each separate step has to be investigated with great care,
the combination of all the steps have in most cases a considerable
overlap, which may serve as a cross-examination of one piece of data
in respect to the other.

2.3.   Determination of Exposed and Buried
Residues in Macromolecules

Besides the direct determination of structure and dynamics within
paramagnetic adducts, paramagnetic enhancement of nuclear relaxation
has been widely exploited for several purposes. It has long been
known as fact that paramagnetic relaxation reagents, by decreasing
spin-lattice relaxation times and by suppressing nuclear Overhauser
effects, are powerful aids in the analysis of ^{13}C NMR spectra [77-80].
Paramagnetic β-diketone derivatives of Cr, Mn, Fe, Ni, and Gd have
been alternatively used, providing the desired T_1 and/or nuclear
Overhauser effects (NOE) at molar ratios as low as 1:500. Whenever
no specific complexation occurs, the observed relaxation could be
interpreted [80] in terms of electron-nuclear dipole-dipole interactions modulated by translational motion and electronic relaxation
[see Sec. 1.2, Eqs. (8) and (9)]. In cases where the paramagnetic
relaxation reagent binds to or orientates the substrate through
steric, electrostatic, or hydrogen-bonding effects, the observed
relaxation also depends on the strength of the interaction.

   Nuclear relaxation investigations of solutions of stable
nitroxide free radicals in proton donor solvents have been extensively carried out as a useful tool for studying molecular motions
and intermolecular interactions in liquids. Since both dipolar and
scalar mechanisms can be effective, either of which being modulated
by more than one type of motion of the interacting species (e.g.,
either translational diffusion of the interacting molecules or rotational diffusion of the associated species can be responsible of
modulation of the dipolar interaction), identification of the relevant

type of motion is not straightforward. In certain cases the dominant
motion could be assumed [81-83], thereby allowing evaluation of corre
lation times, activation energies, and intermolecular distances. A
detailed study of magnetic field dependence of spin-lattice relaxation
rates in aqueous [84] or alcoholic [85] nitroxide solutions has been
suggested for the identification of the differences in spectral density functions attributable to translational or rotational dipolar
mechanisms, the scalar mechanism being usually negligible [86].

The energy level diagram for a nuclei spin I = 1/2 coupled to
an unpaired electron spin S = 1/2 is shown in Figure 9. The scalar
relaxation transition is shown as c, whereas the effective dipolar
relaxation transitions are labeled p, q, r, and s. At low fields

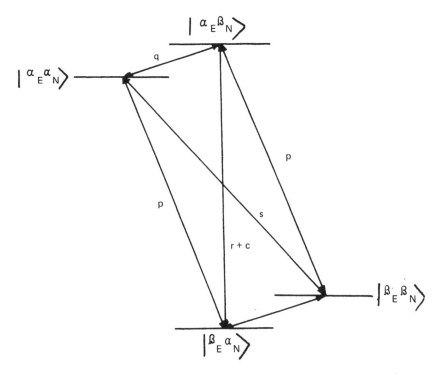

FIG. 9. Energy level diagram for combined spin states for a proton
(I = 1/2) coupled to an electron (S = 1/2); p, q, r, s, and c are
relaxation transitions.

the dipolar components are in the ratio $q/r/s = 3:2:12$. When rotational diffusion is the governing motion, the total dipolar rate
$(= 2q + r + s)$ is given by [11,87]:

$$R_{dr} = \frac{3}{10}\gamma_S^2\gamma_I^2\hbar^2\tau_c d_r^{-6}[n_I N_S N_I^{-1}][J_r(\omega_I) + \frac{7}{3}J_r(\omega_S)] \qquad (26)$$

where $J_r(\omega_I) = 1/(1 + \omega_I^2\tau_c^2)$ and τ_c is the dipolar correlation time for rotational diffusion, d_r is the average pair radius for the rotating species, N_S and N_I represent the electron and nuclear spin concentrations, n_I is the number of equivalent receptor nuclei bound near each electron.

When translational diffusion is the governing motion, the total dipolar rate is given by [17,87]:

$$R_{dt} = \frac{2\pi}{5}\hbar^2\gamma_S^2\gamma_I^2\tau_d d_t^{-3}N_S[J_t(\omega_I) + \frac{7}{3}J_t(\omega_S)] \qquad (27)$$

where

$$J_t(\omega_I) \cong [1 + 0.9(\omega_i\tau_d)^{1/2} + 1.5(\omega_i\tau_d)^{3/2}]^{-1} \qquad (28)$$

Here τ_d is the dipolar correlation time for translational diffusion; d_t is the distance of closest approach ($\tau_d = d_t^2/D$ where D is the diffusion coefficient). One may note that Eqs. (26)-(28) are particular cases of Eq. (8) with different notations.

The predicted spectral density-magnetic field correlation curves are compared in Figure 10 for the two different mechanisms. The theoretical curves are apparently different enough to allow straightforward identification of one dominating motion when the other is not effective at all. However, when a mixture of the two motions is present, introduction of a mixing parameter in the fitting procedure $[R_{obs} = xR_{dr} + (1 - x)R_{dt}]$ brings together a very large uncertainty.

In such cases a simple method has been suggested [87] by which to identify the dominant motional component. The method involves an analysis of relaxation of solvent nuclei in a nitroxide solution as a function of solvent concentration in an inert cosolvent. Since

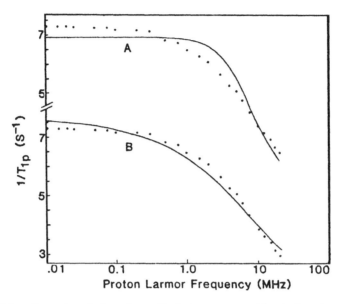

FIG. 10. Experimental and predicted paramagnetic-induced rates on
(a) rotational and (b) translational models alone for data at 278 K.

the R_{dt} component is expected to give only a small contribution to
such concentration dependency, slopes and intercepts of the straight
lines are obtained when the reduced paramagnetic enhancement
$[= (R_{obs} - R_{blank})/N_S]$ is plotted against the reduced molar fraction
of the nitroxide-solvent complex (the formation constant can be
independently measured, e.g., by ESR). The intercept provides
R_{dt}/N_S directly whereas the slope yields $(R_{dr} - R_{dt})$ [87].

The effects likely to govern nuclear spin relaxation process
of solvent molecules in solutions of stable free radicals or metal
complexes can be schematized into five possibilities: (1) no interaction, (2) specific steric or motional effects only, (3) polar or
electrostatic effects, (4) hydrogen-bonding or outer sphere complexation effects, and (5) long-lived chemical combinations. If the fifth
effect is excluded, it may be considered that whenever intermolecular
forces among solvent, solute, and paramagnetic label molecules are of
similar type and strength, paramagnetic centers can be used to delin-
eate the solvent-exposed surface of any cosolute molecule [88-90].

Such method has been successfully applied to peptides in solution [88-95], where delineation of solvent-exposed vs. solvent-shielded protons is of primary relevance for establishing the hydrogen-bonding network of the preferred conformation. The advantage of the method as compared with other NMR methods, including proton-deuterium exchange rates [96], temperature dependence of chemical shifts [97], solvent dependence of chemical shifts [98], pH-rate profiles [99], and solvent saturation [100], is that minimal perturbations of the conformation are involved and that it is applicable to carbon-bound protons as well as to amide protons. For these kinds of studies stable nitroxides or metal ions complexed by strong ligands can be used. The effectiveness of nitroxides as relaxation reagents in proton-accepting solvents reflects the degree of exposure of the observed proton to the external environment, although there is some evidence of preferential hydrogen bonding in DMSO [92]. In contrast, metal complexes show a certain preferential affinity toward some functional groups and their uses should therefore be limited. As an example, the dependence of T_{1p}^{-1} on nitroxide concentration (TEMPO = 2,2,6,6-tetramethyl-piperidineoxyl) is shown in Figure 11 for the NH and H protons of gramicidin S [91]. All the protons are affected by the nitroxide but the size of the experienced relaxation enhancement reflects the respective distances of closest approach between the observed proton and the free radical. As a consequence, the local conformation of gramicidin S governs the extent of relaxation enhancement of any proton; the solvent-shielded Orn H and Leu NH extending toward the interior of the cyclic peptide, are therefore experiencing the lowest relaxation enhancement. Hydrogen bonding was shown to be the driving force for the nitroxide-gramicidin S association, leading to the observed paramagnetic effect [94]; a large downfield shift was in fact detected for the TEMPO methyl protons at increasing concentrations of gramicidin S, whereby an upper limit for the dissociation constant was calculated ($K_D \leqslant 0.5$ mol/dm^3).

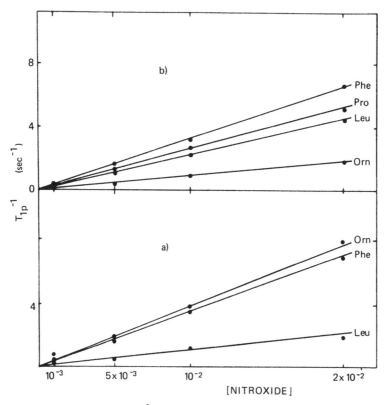

FIG. 11. Dependence of T_{1p}^{-1} on TEMPO concentration for (a) NH protons and (b) H_α protons of gramicidin S 40 mM in DMSO-d_6 at 299 K.

2.4. Paramagnetic Probes of the Antibody Combining Sites

An antibody (immunoglobulin) is a protein synthesized by an organism when stimulated by a foreign antigenic substance. The structural features of the antibody molecule are shown in Figure 12. The common feature shared by all the immunoglobulins is the presence of six globular regions, linked together by short sections of relatively less structured polypeptide chains. The greatest sequence variability between immunoglobulins is confined to the V_H and V_L domain (H and L mean heavy and light, respectively) which contain the antigen binding

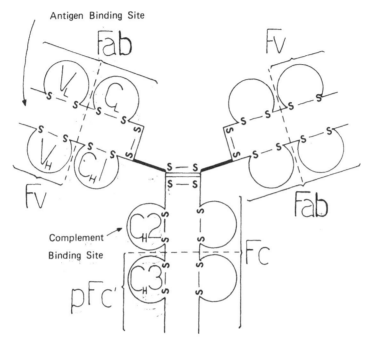

FIG. 12. Structure of IgG. The circled regions show the different domains (V = variable, C = constant, H = belonging to the heavy chain, L = belonging to the light chain). Fc, pFc', Fab, and Fv indicate various proteolytic fragments.

site. The complement binding site is the site for binding of the first protein of the complement sequence (a series of proteins triggered to cause lysis of cells to which antibody is bound).

Burton et al. [58] showed that the nonimmune rabbit IgG possesses two equivalent sites with high affinity toward Gd(III) in the C_H3 domains of the Fc region. From comparison of proton and deuteron relaxation rates of water combined with computer fitting of the frequency dependence, τ_R was fixed within the limits 5-10 nsec. It was therefore suggested that an internal motion was present in the Fc portion of the antibody. However, it was also noticed that the observed internal motion might reflect a local flexibility around the metal ion binding site, without any functional significance.

Most of the subsequent NMR or ESR studies were performed with mouse myeloma IgA protein 315, since it is produced by genetically identical cells, thus yielding an homogeneous population. On the contrary, the usual number of different antibodies produced against a certain antigenic hapten ranges between 3,000 and 30,000. By extensive screening, ligands were discovered for myeloma antibodies; the binding of one of them was studied by x-ray crystallography [101,102] and also by NMR [103,104]. In other studies, Dwek et al. [105,106] focused on the MOPC 315 myeloma antibody directed against dinitrophenyl (DNP) hapten and characterized the MOPC 315-DNP complex by NMR. They showed that it was possible to use the difference between the NMR spectra of the antibody and the antibody-hapten complex to get information about the structure of the binding site. The only resonances appearing in the difference NMR spectrum were in fact those undergoing chemical shift upon binding of the DNP hapten. In previous studies [107] nitroxide spin label had been attached to DNP, leading to observation of proton resonances within a radius of about 15 Å from the unpaired electron in the difference NMR spectrum of the Fv fragment of the antibody with and without the spin-labeled DNP (DNP-SL).

The development of hybridoma technology for the achievement of large amounts of monoclonal antibodies together with selective in vitro incorporation of deuterated amino acids were used by McConnell et al. [108-111], who amplified the early investigations of Dwek et al. [105-107], providing more detailed information about the residues in and around the binding site. The antigen was properly chosen to be a suitable spectroscopic probe of its interaction with the antibody. Its formula is shown in Figure 13, while the method of preparation is described in [108].

The DNP-SL was bound to an anti-DNP-SL monoclonal IgG_1 antibody produced by a hybridoma cell line [108]. It was shown by tryptophan fluorescence quenching that the dissociation constant for the DNP-SL-antibody complex is 2.5×10^{-7} mol/dm^3. Moreover, ESR evidence was collected of a motional correlation time of 2×10^{-8} sec for the bound nitroxide, in agreement with that expected for the Fab molecule [108].

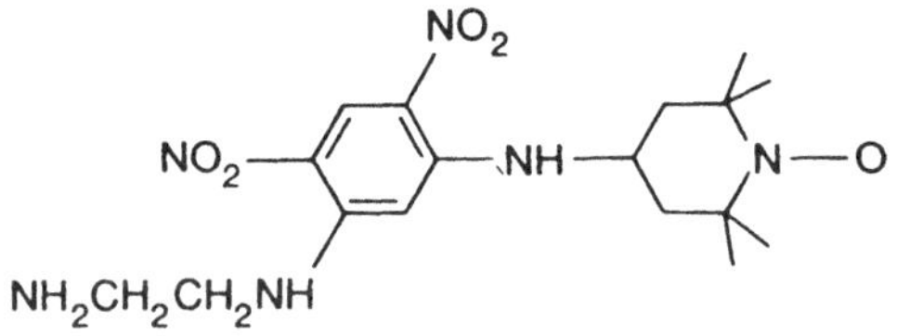

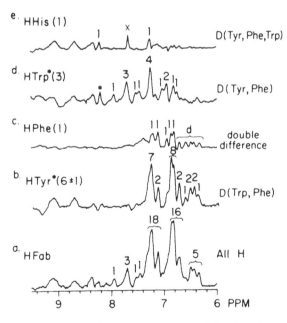

FIG. 13. Spin-label hapten.

When looking at the aromatic region within the ^{1}H NMR difference
spectrum (Fab)-(Fab-DNP-SL) (Fig. 14), it was possible to isolate
the resonance signals for each aromatic amino acid in the combining
site. At this point advantage was taken from the easy adaptability
of the hybridoma cell line to grow on deuterated amino acids or

FIG. 14. Spectrum of Fab minus spectrum of Fab with bound spin-label
hapten bound for various deuterated proteins. Deuterated amino acids
are indicated at the right; major contributions to proton resonances
are at the left, followed by the number of residues inferred. H(Trp)*
and H(Tyr)* contain the two single resonances from histidine. Resonances labeled d are beyond the full broadening range of the spin label
and are called "distant." (Reproduced by permission from [109].)

combinations of deuterated amino acids. The NMR spectrum of histidines (His) was obtained by growing the cell line on a combination
of deuterated tryptophan (Trp), tyrosine (Tyr), and phenylalanine
(Phe). In the same way the NMR spectra of tryptophans and tyrosines
were also obtained (Fig. 14), whereas the spectrum of phenylalanine
was derived from the double difference [(Fab)-(Fab-DNP-SL)]-
[(Fab{D(Phe)})—(Fab{D(Phe)}—DNP-SL)] (Fig. 14). In this case
spectra were measured and difference spectra were calculated for two
preparations of Fab differing in the deuteration of the amino acid
whose contribution was therefore elicited in the double-difference
spectrum. From integration of the spectra it was estimated that one
His, three Trp, one Phe, and six or seven Tyr are present at the
combining site [109]. Since as many as 40 amino acids are expected
to occupy the spherical volume within a radius of 15 Å from the
unpaired electron, the 11 or 12 identified aromatic amino acids are
only a fraction of the amino acids affected by the electron spin.
In fact, in the same way, three to five threonines, four valines,
three leucines, one isoleucine, and four to six alanines were identified, constituting, together with the aromatic amino acids, about
80% of all the amino acids at the combining site. Among them one
Trp and two Tyr residues were shown to be in close contact with the
aromatic moiety of the hapten [109] since corresponding chemically
shifted resonances were appearing in the difference spectrum (Fab)-
(Fab-DNP-gly). The Trp residue was also shown to be responsible for
the charge transfer interaction affecting the UV spectrum of DNP-SL
[109].

The hapten and antibody are in dynamic equilibrium as shown by
rapid exchange of hapten molecules between bound and nonbound sites
[109]. By applying the Solomon-Bloembergen approach [Eqs. (6) and
(7), Sec. 1.2], it was calculated [110] that the line broadening
when the binding site is fully occupied is given by:

$$\Delta \nu = \frac{<(30.1/r)^6>}{\pi} \tag{29}$$

Here the parentheses designate an average over intramolecular motions and the constant derived from substitution of fundamental constants in the SB equation ($\tau_c = 1.5 \times 10^{-8}$ was taken as an estimate of molecular tumbling of any Fab fragment). In the case of any fractional occupancy f, the line broadening is determined by $f \, \Delta\nu$. The intensity $I(\omega)$ of a Lorentzian proton resonance line is given by:

$$I(\omega) = \frac{1}{\pi} \frac{T_2}{1 + (\omega - \omega_0)^2 T_2^2} \tag{30}$$

Under conditions of fast exchange the peak height in the difference spectrum $I(f, \omega_0)$ is given by [110]:

$$I(f, \omega_0) = \frac{1}{\pi} \frac{T_{2M} T_{2N}}{f \, \Delta T_2 + T_{2M}} \tag{31}$$

where $\Delta T_2 = T_{2N} - T_{2M}$, T_{2N} is the transverse relaxation time of the proton in the Fab without DNP-SL, T_{2M} is the transverse relaxation time of the same proton in the Fab with DNP-SL, and $1/T_{2M} = 1/T_{2N} + <(30.1/r)^6>$. Under intermediate exchange conditions, the peak height is also dependent on the off-rate constant k:

$$I(k, f, \omega_0) = \frac{1}{\pi} \frac{[T_{2M}(1 - f) + \Delta T_2 (1 - f)^2 + k T_{2N} T_{2M}]}{[(1 - f) + \Delta T_2 k f + T_{2M} k]} \tag{32}$$

Equation (32) reduces to Eq. (31) when k is large, whereas in the slow exchange limit, k = 0, one obtains:

$$I(0, f, \omega_0) = \frac{1}{\pi} [T_{2M} + \Delta T_2 (1 - f)] \tag{33}$$

In order to calculate distances and exchange rates it was shown that normalized intensities J(f) are more convenient [110]:

$$J(f) = \frac{[I(k, 0, \omega_0) - I(k, f, \omega_0)]}{[I(k, 0, \omega_0) - I(k, 1, \omega_0)]} \tag{34}$$

Equations including eventual changes in chemical shift due to hapten binding were also considered [112].

By feeding the cells with deuterated tryptophan and phenylalanine and also with tyrosine partially deuterated in positions 2
and 6 (meta to the -OH group), it was possible to remove J splittings
and the major sources of diamagnetic relaxation for tyrosine protons
3 and 5 (ortho to the -OH group), which were showing as sharp and
well-resolved singlets in the difference spectrum (Fab)-(Fab-DNP-SL)
[110]. By comparing difference spectra of the type Fab(x% DNP-Gly)-
(Fab-DNP-SL) it was therefore possible to follow the changes in
chemical shift that different protons undergo upon binding of the
hapten. The position and linewidths of selected lines in the difference spectra could be accounted for only by definite numerical values
of k which could therefore be estimated at 500 sec^{-1} for the DNP-Gly
hapten.

When titrating the Fab with the DNP-SL hapten, it was found
that the peak height in the difference spectrum was independent on
the electron-nucleus distance provided the change in chemical shift
is large, while it depends only on the off rate and the transverse
relaxation time T_2. Under these circumstances the measurement of
J(f) yielded an estimate of k for the spin label hapten itself as
shown in Figure 15. The experimental and theoretical J(f) data for
proton signals from either H-2,6-Tyr [Fig. 15(a)] or H-3,5-Tyr [Fig.
15(b)] lead to the conclusion that k is 350 sec^{-1} for DNP-SL at 35°C
[110]. Knowledge of k allowed the authors to calculate distances
for tyrosine protons, as reported in Table 1. By the same way, one
-CH$_3$ group of methionine and a ε-CH$_2$ group of lysine were calculated
at 12 ± 0.75 Å from the impaired electron. However, it can be
noticed that the method is only accurate in measuring distances
longer than 8 Å, a problem which should be overcome by using haptens
with larger off rates.

Distances between some amino acids to a paramagnetic center
cannot by themselves provide a three-dimensional structure of the
combining site. However, by computer modeling using the known amino
acid sequences and by comparison with homologous antibodies whose
structure have been solved by x-ray crystallography, such information
can be obtained.

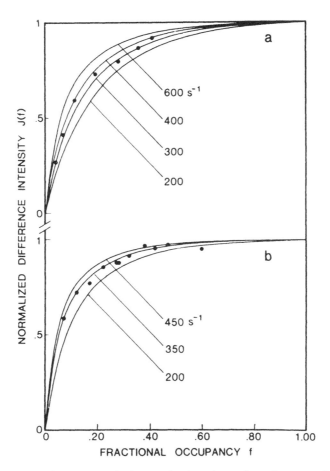

FIG. 15. Experimental and theoretical values for the normalized difference spectra peak intensities J(f) for a proton signal from H-2,6-Tyr (upper part a) and a proton signal from H-3,5-Tyr (lower part b). The solid lines are from theoretical calculations using the indicated off rates k for the spin-label hapten and changes in chemical shifts and linewidths given in Table 1. (Reproduced by permission from [110].)

TABLE 1

Calculated Distances and Relevant Data

	Chemical shift (Hz)[a]	Linewidth $\Delta\nu_{\frac{1}{2}}$ (Hz)	Distance, r (Å)
H-3,5-Tyr			
A[b]	-18.65	7.6	14 ± 1.5
B (2 pairs)	0	10.1[c]	<9.5[c]
C	0	5.2	11 ± 2
D	7	7	12 ± 2
E	3	5	13 ± 1.5
F	25	10.5	≤10
G	-200	6.3	?
H	-38.65	7.2	<12
H-2,6-Tyr			
A	?		
B	Overlap		
C	2.5	5	12 ± 1
D	0	8.5	13 ± 11
E	23.5	10	>11
F	7.5	14	12 ± 1
G			"Far"
H	195	13	?

[a]Positive change in chemical shift is toward lower field.
[b]Letters designate proton resonances.
[c]This line may be inhomogeneous because it arises from two pairs of protons. If the width is assumed to be 5 Hz, the calculated distance is 11 Å.
Source: Reprinted by permission from [110].

ABBREVIATIONS

acac   acetylacetonate

dpm   2,2,6,6-tetramethylheptanedione

EPR   electron paramagnetic resonance

NMR   nuclear magnetic resonance

NOE   nuclear Overhauser effect

SB   relaxation theory as developed by Solomon and Bloembergen

ZFS   zero field splitting

REFERENCES

1. T. J. Swift and R. E. Connick, *J. Chem. Phys., 37,* 307 (1962).

2. Z. Luz and S. Meiboom, *J. Chem. Phys., 40,* 2686 (1964).

3. J. S. Leigh, *J. Magn. Reson., 4,* 308 (1971).

4. Y. F. Lam, G. P. P. Kuntz, and G. Kotowycz, *J. Am. Chem. Soc., 96,* 1834 (1974).

5. G. P. P. Kuntz and G. Kotowycz, *Biochemistry, 14,* 4144 (1975).

6. R. A. Dwek, *Nuclear Magnetic Resonance in Biochemistry,* Oxford University Press, New York, 1973.

7. D. R. Burton, S. Forsén, G. Karlstrom, and R. A. Dwek, *Prog. NMR Spectrosc., 13,* 1 (1980).

8. N. Niccolai, E. Tiezzi, and G. Valensin, *Chem. Rev., 82,* 359 (1982).

9. J. Kowalewski, L. Nordenskiöld, N. Benetis, and P. O. Westlund, *Prog. NMR Spectrosc., 17,* 141 (1985).

10. I. Bertini and R. S. Drago (eds.), *ESR and NMR of Paramagnetic Species in Biological and Related Systems,* D. Reidel, Dordrecht, 1979.

11. I. Solomon, *Phys. Rev., 99,* 559 (1955).

12. I. Solomon and N. Bloembergen, *J. Chem. Phys., 25,* 261 (1956).

13. N. Bloembergen, *J. Chem. Phys., 27,* 572 (1957).

14. A. Abragam, *The Principles of Nuclear Magnetism,* Clarendon Press, Oxford, 1961.

15. R. E. Connick and D. Fiat, *J. Chem. Phys., 44,* 4103 (1966).

16. J. Reuben, G. H. Reed, and M. Cohn, *J. Chem. Phys., 52,* 1617 (1970).

17. H. Pfeifer, *Ann. Phys.*, *8*, 1 (1961).

18. H. Pfeifer, *Z. Naturforsch.*, *17*, 279 (1962).

19. J. Bloch and G. Navon, *J. Inorg. Nucl. Chem.*, *42*, 693 (1980).

20. J. Kärger and H. Pfeifer, *Ann. Phys.*, *22*, 51 (1980).

21. H. Sternlicht, *J. Chem. Phys.*, *43*, 2250 (1965).

22. I. Bertini, C. Luchinat, M. Mancini, and G. Spina, *J. Magn. Reson.*, *59*, 213 (1984).

23. H. A. Bergen, R. M. Golding, and L. C. Stubbs, *Mol. Phys.*, *37*, 1371 (1979).

24. I. Bertini, F. Briganti, C. Luchinat, M. Mancini, and G. Spina, *J. Magn. Reson.*, *63*, 41 (1985).

25. A. Abragam and B. Bleaney, *Electron Paramagnetic Resonance of Transition Ions*, Oxford University Press, Oxford, 1970.

26. D. Waysbort and G. Navon, *Chem. Phys.*, *49*, 333 (1980).

27. N. Bloembergen and L. O. Morgan, *J. Chem. Phys.*, *4*, 842 (1961).

28. M. Rubinstein, A. Baram, and Z. Luz, *Mol. Phys.*, *20*, 67 (1971).

29. S. H. Koenig, *J. Magn. Reson.*, *31*, 1 (1978).

30. D. M. Doddrell, D. T. Pegg, and M. R. Bendall, *Chem. Phys. Lett.*, *40*, 142 (1976).

31. D. M. Doddrell, M. R. Bendall, and A. K. Gregson, *Aust. J. Chem.*, *29*, 55 (1976).

32. D. T. Pegg and D. M. Doddrell, *Aust. J. Chem.*, *29*, 1869 (1976).

33. D. T. Pegg, D. M. Doddrell, M. R. Bendall, and A. K. Gregson, *Aust. J. Chem.*, *29*, 1885 (1976).

34. H. L. Friedman, M. Holz, and H. G. Hertz, *J. Chem. Phys.*, *70*, 369 (1979).

35. N. Benetis, J. Kowalewski, L. Nordenskiöld, H. Wennerstrom, and P. O. Westlund, *Mol. Phys.*, *50*, 515 (1983).

36. P. O. Westlund, H. Wennerstrom, L. Nordenskiöld, J. Kowalewski, and N. Benetis, *J. Magn. Reson.*, *59*, 91 (1984).

37. N. Benetis, J. Kowalewski, L. Nordenskiöld, and U. Edlund, *J. Magn. Reson.*, *58*, 282 (1984).

38. M. Gueron, *J. Magn. Reson.*, *19*, 58 (1975).

39. A. J. Vega and D. Fiat, *Mol. Phys.*, *1*, 347 (1976).

40. M. E. Johnson, L. W.-M. Fung, and C. Ho, *J. Am. Chem. Soc.*, *99*, 1245 (1977).

41. G. N. La Mar, J. T. Jackson, and R. G. Bartsch, *J. Am. Chem. Soc.*, *103*, 4405 (1981).

42. D. J. Livingston, G. N. La Mar, and W. D. Brown, *Science, 220,* 71 (1983).

43. D. Waysbort and G. Navon, *J. Chem. Phys., 62,* 1021 (1975).

44. D. Waysbort and G. Navon, *Chem. Phys., 28,* 83 (1978).

45. D. Waysbort and G. Navon, *J. Chem. Phys., 68,* 3074 (1978).

46. H. P. W. Gottlieb, M. Barfield, and D. M. Doddrell, *J. Chem. Phys., 67,* 3785 (1977).

47. L. Nordenskiöld, A. Laaksonen, and J. Kowalewski, *J. Am. Chem. Soc., 104,* 379 (1982).

48. J. Kowalewski, A. Laaksonen, L. Nordenskiöld, and V. R. Saunders, *J. Magn. Reson., 53,* 346 (1983).

49. P. J. Andree, *J. Magn. Reson., 29,* 419 (1978).

50. B. D. Sykes, W. E. Hull, and G. H. Snyder, *Biophys. J., 21,* 137 (1978).

51. J. Granot, *J. Magn. Reson., 49,* 257 (1982).

52. E. Sletten, J. T. Jackson, P. D. Burns, and G. N. La Mar, *J. Magn. Res., 52,* 492 (1983).

53. J. Eisinger, R. G. Shulman, and W. E. Blumberg, *Nature, 192,* 963 (1961).

54. J. Eisinger, R. G. Shulman, and B. M. Szymanski, *J. Chem. Phys., 336,* 1721 (1962).

55. A. S. Mildvan and M. Cohn, *Biochemistry, 2,* 910 (1963).

56. G. Navon, *Chem. Phys. Lett., 7,* 390 (1970).

57. T. Kushnir and G. Navon, *J. Magn. Reson., 56,* 373 (1984).

58. D. R. Burton, R. A. Dwek, S. Forsén, and G. Karlstrom, *Biochemistry, 16,* 250 (1977).

59. A. S. Mildvan and M. Cohn, *Biochemistry, 6,* 1805 (1967).

60. G. Navon, R. G. Shulman, J. Wyluda, and T. Yamane, *Proc. Natl. Acad. Sci. USA, 60,* 86 (1968).

61. Z. Luz and S. Meiboom, *J. Chem. Phys., 40,* 2686 (1964).

62. A. Lanir and G. Navon, *Biochemistry, 11,* 3536 (1972).

63. A. S. Mildvan, *Phil. Trans. R. Soc. Lond. B, 293,* 65 (1981).

64. A. S. Mildvan, E. T. Kaiser, P. R. Rosevear, and H. N. Bramson, *Fed. Proc., 43,* 2634 (1984).

65. M. L. De Pamphilis and W. W. Cleland, *Biochemistry, 12,* 3714 (1973).

66. R. D. Cornelius and W. W. Cleland, *Biochemistry, 17,* 3279 (1978).

67. W. W. Cleland and A. S. Mildvan, *Adv. Inorg. Biochem., 1,* 163 (1979).

68. J. Granot, H. Kondo, R. N. Armstrong, A. S. Mildvan, and E. T. Kaiser, *Biochemistry, 18,* 2339 (1979).

69. J. Granot, A. S. Mildvan, and E. T. Kaiser, *Arch. Biochem. Biophys., 205,* 1 (1980).

70. J. Granot, A. S. Mildvan, H. N. Bramson, and E. T. Kaiser, *Biochemistry, 19,* 3537 (1980).

71. J. Granot, K. J. Gibson, R. L. Switzer, and A. S. Mildvan, *J. Biol. Chem., 255,* 10931 (1980).

72. R. K. Gupta, C. H. Fung, and A. S. Mildvan, *J. Biol. Chem., 251,* 2421 (1976).

73. R. K. Gupta and A. S. Mildvan, *J. Biol. Chem., 252,* 5967 (1977).

74. R. K. Gupta and J. L. Benovic, *J. Biol. Chem., 253,* 8878 (1978).

75. J. Reuben and F. J. Kayne, *J. Biol. Chem., 246,* 6227 (1971).

76. A. S. Mildvan, *Adv. Enzymol. Relat. Areas Mol. Biol., 49,* 103 (1979).

77. G. N. La Mar, *Chem. Phys. Lett., 10,* 230 (1971).

78. R. Freeman, K. G. R. Packler, and G. N. La Mar, *J. Chem. Phys., 55,* 4586 (1971).

79. D. F. S. Natusch, *J. Am. Chem. Soc., 93,* 2566 (1971).

80. G. C. Levy, U. Edlund, and C. E. Holloway, *J. Magn. Reson., 24,* 375 (1976) and references therein.

81. K. Endo, I. Morishima, and T. Yonezawa, *J. Chem. Phys., 67,* 4760 (1977).

82. A. G. Goetz, D. Z. Denney, and J. A. Potenza, *J. Phys. Chem., 83,* 3029 (1979).

83. N. Chandrakumar and P. T. Narasimhan, *Mol. Phys., 45,* 179 (1982).

84. B. Borah and R. G. Bryant, *J. Chem. Phys., 75,* 3297 (1981).

85. H.-W. Nientiedt, K. Bundfuss, and W. Muller-Warmuth, *J. Magn. Reson., 43,* 154 (1981).

86. K. Meise, W. Muller-Warmuth, and H.-W. Nientiedt, *Ber. Bunsenges. Phys. Chem., 80,* 584 (1976).

87. J.-A. K. Bonesteel, B. Borah, and R. D. Bates, Jr., *J. Phys. Chem., 88,* 2141 (1984).

88. K. D. Kopple and T. J. Schamper, *J. Am. Chem. Soc., 94,* 3644 (1972).

89. K. D. Kopple and T. J. Schamper, *J. Am. Chem. Soc., 99,* 7698 (1977).

90. N Niccolai, N. Zhou, C. Rossi, P. Mascagni, and W. A. Gibbons, in *Peptides, Synthesis-Structure-Function,* Proc. 7th Am. Peptide Symp. (D. H. Rich and E. Gross, eds.), pp. 307-310, Pierce Chemical Co., Rockford, Ill., 1981.

91. N. Niccolai, G. Valensin, C. Rossi, and W. A. Gibbons, *J. Am. Chem. Soc.*, *104*, 1534 (1982).

92. K. D. Kopple and P.-P. Zhu, *J. Am. Chem. Soc.*, *105*, 7742 (1983).

93. K. D. Kopple, *Int. J. Peptide Protein Res.*, *21*, 43 (1983).

94. N. Niccolai, C. Rossi, G. Valensin, P. Mascagni, and W. A. Gibbons, *J. Phys. Chem.*, *88*, 5689 (1984).

95. N. Zhou, P. Mascagni, W. A. Gibbons, N. Niccolai, C. Rossi, and H. Wyssbrod, *JCS Perkin II*, 581 (1984).

96. A. Stern, W. A. Gibbons, and L. C. Craig, *Proc. Natl. Acad. Sci. USA*, *61*, 734 (1968).

97. M. Ohnishi and D. W. Urry, *Biochem. Biophys. Res. Commun.*, *36*, 94 (1969).

98. T. P. Pitner and D. W. Urry, *J. Am. Chem. Soc.*, *94*, 1399 (1972).

99. R. S. Molday, S. W. Englander, and R. G. Kallen, *Biochemistry*, *11*, 150 (1970).

100. T. P. Pitner, J. D. Glickson, R. Rowan, J. Dadok, and A. A. Bothner-By, *J. Am. Chem. Soc.*, *97*, 5917 (1975).

101. L. H. Amzel, R. J. Poljak, F. Saul, J. M. Varga, and F. F. Richards, *Proc. Natl. Acad. Sci. USA*, *71*, 1427 (1974).

102. E. A. Padlan, D. R. Davies, S. Rudikoff, and M. Potter, *Immunochemistry*, *13*, 945 (1976).

103. A. M. Goetze and J. H. Richards, *Biochemistry*, *17*, 1733 (1978).

104. D. A. Kooistra and J. H. Richards, *Biochemistry*, *17*, 345 (1978).

105. R. A. Dwek, S. Wainhobson, S. Dower, P. Getting, B. Sutter, S. J. Perkins, and D. Givol, *Nature*, *266*, 31 (1977).

106. S. Dower and R. A. Dwek, in *Biological Applications of Magnetic Resonance* (R. G. Shulman, ed.), Academic Press, New York, 1979, pp. 271-303.

107. R. A. Dwek, J. C. A. Knot, D. Marsh, A. C. McLaughlin, and E. M. Press, *Eur. J. Biochem.*, *53*, 25 (1975).

108. K. Balakrishnan, F. J. Hsu, D. G. Hafeman, and H. M. McConnell, *Biochim. Biophys. Acta*, *721*, 30 (1982).

109. J. Anglister, T. Frey, and H. M. McConnell, *Biochemistry*, *23*, 1138 (1984).

110. J. Anglister, T. Frey, and H. M. McConnell, *Biochemistry*, *23*, 5372 (1984).

111. T. Frey, J. Anglister, and H. M. McConnell, *Biochemistry*, *23*, 6470 (1984).

112. R. W. Weine, J. D. Morrisett, and H. M. McConnell, *Biochemistry*, *11*, 3707 (1972).

2

Nuclear Relaxation in NMR of Paramagnetic Systems

Ivano Bertini, Claudio Luchinat, and Luigi Messori
Department of Chemistry
University of Florence
via Gino Capponi 7
I-50121 Florence, Italy

47

1. NUCLEAR RELAXATION AND OBSERVABILITY
OF NMR SIGNALS

Unpaired electrons have relatively large magnetic moments and are
capable of delocalizing directly at or near atomic nuclei in a given
molecule. Finally, they can polarize the two paired electrons in a
different molecular orbital (MO) of the same molecular species in
such a way that the spatial delocalization of each of the two becomes
different; it follows that in every point within a paramagnetic molecule there is a net spin density [1]. Electron spin density causes
large shielding or deshielding of the applied magnetic field at the
resonating nuclei. The shift contribution to the NMR signal of a
given nucleus due to the presence of unpaired lectrons is called
isotropic shift (see Sec. 2.1) when measured under fast rotating
conditions (typically for molecular tumbling rates larger than
10^5 sec^{-1}).

     The presence of unpaired electrons enhances the shift separa-
tion among the various signals. This is a contribution to the sensitivity of the technique with respect to the detection of subtle
structural differences. However, unpaired electrons also cause a
broadening of the signals and therefore a decrease in sensitivity.
Electrons change their M_S values with respect to the quantization
axis with characteristic time constants T_1 in the range 10^{-7}-10^{-12}
sec, which are much shorter than the corresponding times in the
nuclear case. When a magnetically active nucleus feels unpaired
electrons, several mechanisms become operative which lead to a
decrease in its nuclear relaxation times. One of these is the
dipolar coupling: under certain conditions it holds that the more
slowly the electron relaxes, the larger is the coupling with the

nucleus and the faster the nucleus relaxes. The qualitative pattern
for the contact coupling is similar.

The linewidth of an NMR (or EPR) signal depends on the transverse relaxation rate (T_2^{-1}) according to the relationship

$$T_2^{-1} = \pi \Delta \nu_{1/2} \tag{1}$$

where $\Delta \nu_{1/2}$ is the halfheight linewidth. The transverse relaxation
rate is defined as the rate constant for the first-order process
which brings the magnetization in the xy plane, i.e., the plane
orthogonal to the applied magnetic field, to its equilibrium value
from any nonequilibrium state. The equilibrium value of the xy
magnetization is zero and goes to a finite value different from zero
during the resonance experiment. This definition is similar to that
of T_1^{-1} (longitudinal relaxation rate) which is referred to the magnetization along the z axis; the latter has a finite value at equilibrium which depends on the magnetic field strength.

Unpaired electrons affect the nuclear relaxation rates and
therefore the linewidth of a NMR signal is bound to the electronic
relaxation rate. In Table 1 the latter values are reported for a
series of metal ions together with the nuclear halfheight linewidth
values calculated for a proton 500 pm away from the metal on the
assumption that only the dipolar mechanism contributes to the broadening. A signal of 5,000 Hz of linewidth is 50 ppm broad at 100 MHz
and is hardly detectable on a few hundreds ppm scale (spectral width).
From Table 1 it appears that low-spin iron(III) does not provide
severe broadening of the resonance line, just like the lanthanides(III)
(except Gd^{3+}) and high-spin cobalt(II). Their compounds can be properly investigated through NMR under any circumstance. In copper(II),
manganese(II), and gadolinium(III) compounds, if we restrict our
analysis to slow-rotating molecules (see Secs. 3.1-3.4), the linewidths are so broad that the signals are not observed at all. In
this chapter the factors affecting nuclear relaxation will be analyzed in order to meet the conditions of observability of the signals
and, provided that the signal is observed, the wealth of structural
and dynamic information which can be gained will be reviewed.

TABLE 1

Electronic Relaxation Rates for Various Metal Ions

Metal ion	τ_s^{-1} (sec^{-1})	Nuclear line broadening (Hz)[a]
Ti^{3+}	10^9-10^{10}	3,000-500
VO^{2+}	10^8-10^9	20,000-3,000
V^{3+}	2×10^{11}	100
V^{2+}	2×10^9	9,000
Cr^{3+}	2×10^9	9,000
Cr^{2+}	10^{11}	300
Mn^{3+}	10^{10}-10^{11}	3,000-300
Mn^{2+}	10^8-10^9	200,000-40,000
Fe^{3+} (H.S.)	10^{10}-10^{11}	5,000-400
Fe^{3+} (L.S.)	10^{11}-10^{12}	40-10
Fe^{2+} (H.S.)	10^{12}	70
Co^{2+} (H.S.)	10^{11}-10^{12}	200-50
Co^{2+} (L.S.)	10^9-10^{10}	3,000-500
Ni^{2+}	10^{10}-10^{12}	1,000-25
Cu^{2+}	3×10^8-10^9	9,000-3,000
Ru^{3+}	10^{11}-10^{12}	40-10
Re^{3+}	10^{11}	100
Gd^{3+}	10^8-10^9	400,000-60,000
Dy^{3+}	1.25×10^{12}	100
Ho^{3+}	1.35×10^{12}	100
Tb^{3+}	1.25×10^{12}	100
Tm^{3+}	1.25×10^{12}	70
Yb^{3+}	10^{12}	30

[a]For ^1H, dipolar relaxation only, r = 500 pm, B_0 = 2.35T.
Source: Reproduced with permission from [1].

2. PHYSICAL CONSEQUENCES OF UNPAIRED ELECTRON-NUCLEUS COUPLING

2.1. The Isotropic Shift

Although the analysis of the isotropic shifts is outside the aim of this chapter, an understanding of it is instructive since it arises from the coupling between unpaired electrons and nuclei. A contribution to the isotropic shift comes from the presence of unpaired spin density at the resonating nucleus. It is named contact shift and it is given by:

$$\left(\frac{\Delta\nu}{\nu_0}\right)^{con} = \frac{A_c}{\hbar\gamma_N B_0} \langle S_z\rangle \tag{2}$$

where $\Delta\nu$ is the isotropic shift in Hz, ν_0 the reference frequency, A_c the hyperfine coupling constant, $\hbar$ the Planck constant, γ_N the nuclear magnetogyric ratio, B_0 the external magnetic field, and $\langle S_z\rangle$ the expectation value of S_z. A resultant $\langle S_z\rangle$ value different from zero comes from the Boltzmann population of the various M_S levels. This gives the temperature dependence of the contact shift. At infinite temperature all the M_S levels are equally populated and $\langle S_z\rangle = 0$. For an S multiplet,

$$\langle S_z\rangle = -S(S + 1) \frac{\bar{g}\mu_B B_0}{3kT} \tag{3}$$

where $\bar{g}$ is the average g value of the S manifold, μ_B is the Bohr magneton, and k is the Boltzmann constant.

Besides the unpaired spin density at the nucleus, there is unpaired spin density all over the molecule. If the latter kind of spin density gives rise to a magnetic moment which is orientation-dependent, as happens every time there is orbital contribution to the magnetic moment, then there is another contribution to the isotropic shift which does not average to zero upon rotation. Such shift is dipolar in nature, since the coupling occurs through space and attenuates with the third power of the electron-nucleus distance.

Knowledge of the distribution of unpaired spin density all over a molecule from first principles is a difficult task. A common procedure is to divide the dipolar shift into two contributions: one is due to the coupling with electrons localized on the metal (metal-centered dipolar shift), and the other is due to the coupling with unpaired spin density on non-s orbitals of the resonating and directly bound atoms. The latter is known as ligand-centered dipolar shift and sometimes can be estimated from the knowledge of the contact shifts of the nuclei of a molecule [2-4]. Finally, there are some estimates of the error made in considering the electrons on the metal as point dipoles instead of delocalized within the cage formed by the donor atoms [5].

The metal centered dipolar shift is related to the magnetic anisotropy through the following equation:

$$\left(\frac{\Delta\nu}{\nu_0}\right)^{dip} = \frac{1}{4\pi 2r^3}\left[(3\cos^2\theta - 1)\left(\frac{2}{3}\chi_{zz} - \frac{1}{3}\chi_{xx} - \frac{1}{3}\chi_{yy}\right)\right.$$
$$\left. + \sin^2\theta\cos 2\Omega \ (\chi_{xx} - \chi_{yy})\right] \tag{4}$$

where r, θ, and Ω are the polar coordinates of the resonating nucleus with respect to the metal and the molecular axes, and the χ's are the principal values of the magnetic susceptibility tensor [6]. This equation, or the corresponding where χ is substituted by:

$$\chi_{\alpha\alpha} = \mu_0 \frac{g_{\alpha\alpha}^2\mu_B^2 S(S + 1)}{3kT} \qquad (\alpha = x,y,z) \tag{5}$$

where μ_0 is the permeability of vacuum, have been widely used to obtain structural information in solution and constitute the theoretical guidelines for the use of metal ions as shift reagents [1,7-9].

2.2. T_{1M}^{-1} and T_{2M}^{-1}

As mentioned in the introduction, fast-relaxing unpaired electrons provide further efficient pathways for the nucleus to relax. Indeed, since the nucleus relaxes through coupling with fluctuating magnetic fields, it can easily relax through coupling with the electrons which change their orientation with respect to the external magnetic field. Furthermore, if both the resonating nucleus and the unpaired electrons whose magnetic moments are, let us say, aligned along the external magnetic field rotate within the rigid frame of a molecule, the nucleus can easily undergo a transition between two M_I levels. This mechanism is dominant when a molecule rotates fast compared with the electron relaxation rate. Another mechanism is provided by chemical exchange: if the resonating nucleus belongs to a moiety which approaches another moiety with unpaired electrons, binds to it, and then goes away, it feels a fluctuating magnetic field and can therefore relax. This mechanism is operative when the exchange rate is larger than both the electronic relaxation rate and the tumbling rate. The nuclear relaxation rate enhancements due to mechanisms involving unpaired electrons are referred to as T_{1M}^{-1} and T_{2M}^{-1}.

Both nuclear T_1 and T_2 processes require energy exchange between the nuclear spin system and the lattice.

A general treatment has been developed for the case in which the nucleus exchanges energy with the unpaired electrons and then the latter exchange energy with the lattice. This treatment requires that the electron spin system always be in thermal equilibrium with the lattice. In other words, it is assumed that there are very efficient mechanisms which allow the electron to relax. We say that the correlation constant τ_v, which describes the time-dependent interactions between the electrons and the lattice, is much shorter than the electronic relaxation times. These conditions define the Redfield limit [10,11]. Under these circumstances, we define a correlation time τ_c for the time-dependent interactions between the nucleus and

the unpaired electrons. The time τ_c, analogous to τ_v, is a time
constant for the process according to which an ensemble of nuclei
undergo a change in the experienced magnetic field. Again T_{1M}^{-1} and
T_{2M}^{-1} have to be, and indeed are, much smaller than τ_c^{-1} in order to
fulfill the Redfield limit. The mechanisms which determine T_{1M}^{-1} and
T_{2M}^{-1} and which involve electron relaxation are contact and dipolar in
origin just like in the case of the isotropic shift. While the contact contribution depends on the spin density at the nucleus, the
dipolar contribution arises from the unpaired spin density all over
the molecule. The extent of the coupling depends on the square of
the interaction energy. In the case of dipolar coupling it depends
on the square product of the magnetic moments and on the inverse of
the sixth power of the distance. The main dipolar contributions
come from the unpaired electrons on the metal ion and from the spin
density in a non-s orbital (let us say p) of the atom whose nucleus
experiences the resonance phenomenon. The theory is well developed
for the metal-centered dipolar (point dipole) mechanism; there are
also some estimates of the effect of unpaired spin density in a
carbon p_π orbital on the carbon nucleus [2]. We will limit our
interest here to the metal-centered point dipolar coupling which is
presumably the main contribution in ^{1}H NMR spectroscopy.

According to the discussion at the beginning of this section,
the correlation rate τ_c^{-1} for the dipolar coupling is given by:

$$\tau_c^{-1} = \tau_s^{-1} + \tau_r^{-1} + \tau_M^{-1} \tag{6}$$

where τ_s^{-1} is the electronic relaxation rate, τ_r^{-1} is the rotational
correlation rate, and τ_M^{-1} is the exchange rate, if any. In other
words, the fastest of the three mechanisms determines the correlation
rate. Values typical for the three processes are reported in Table 2.

The relationship between $T_{1,2M}^{-1}$ and τ_c through the Zeeman energy
(both nuclear ω_I and electronic ω_S) for an S manifold is given by
Solomon's equations [13]:

TABLE 2

Range of Values for τ_s^{-1}, τ_r^{-1}, and τ_M^{-1}

τ_s^{-1} [a]	10^8-10^{12} sec^{-1}
τ_r^{-1} [b]	10^{11} sec^{-1} for hexaaqua ions up to 10^5 sec^{-1} for 10^5 MW molecules
τ_M^{-1}	10^7 sec^{-1} up to 10^{-3} sec^{-1} for water exchange

[a]See Table 1.
[b]The Stokes-Einstein equation [12] allows an estimate of τ_r according to the relation $\tau_r = 4\pi\eta a^3/3kT$, where η is the viscosity of the solvent and a is the radius of the molecule.

$$T_{1M}^{-1} = \frac{1}{10} \left(\frac{\mu_0}{4\pi}\right)^2 \frac{\hbar^2 \gamma_N^2 \gamma_S^2}{r^6} \left[\frac{\tau_c}{1 + (\omega_I - \omega_s)^2 \tau_c^2} + \frac{3\tau_c}{1 + \omega_I^2 \tau_c^2} \right.$$

$$\left. + \frac{6\tau_c}{1 + (\omega_I + \omega_s)^2 \tau_c^2} \right] \tag{7}$$

$$T_{1M}^{-1} = \frac{2}{15} \left(\frac{\mu_0}{4\pi}\right)^2 \frac{\gamma_N^2 g_e^2 \mu_B^2 S(S + 1)}{r^6} \left[4\tau_c + \frac{\tau_c}{1 + (\omega_I - \omega_s)^2 \tau_c^2} \right.$$

$$\left. + \frac{3\tau_c}{1 + \omega_I^2 \tau_c^2} + \frac{6\tau_c}{1 + (\omega_I + \omega_s)^2 \tau_c^2} + \frac{6\tau_c}{1 + \omega_s^2 \tau_c^2} \right] \tag{8}$$

which were derived in the assumption of an isotropic electron Zeeman interaction. These equations are widely used for mapping a ligand with respect to a metal ion. It should be recalled that in case of chemical exchange, τ_M^{-1} does not presumably affect τ_c^{-1}, but may affect the experimental T_{1p}^{-1} or T_{2p}^{-1} values, i.e., the experimental nuclear relaxation rate enhancements:

$$T_{1p}^{-1} = f_M (T_{1M} + \tau_M)^{-1} \tag{9}$$

$$T_{2p}^{-1} = \frac{f_M}{\tau_M} \frac{T_{2M}^{-2} + T_{2M}^{-1}\tau_M^{-1} + (\Delta\omega_M)^2}{(T_{2M}^{-1} + \tau_M^{-1})^2 + (\Delta\omega_M)^2} \tag{10}$$

where $\Delta\omega_M = 2\pi\ \Delta\nu$ is the isotropic shift in rads sec^{-1} and f_M is the molar fraction of nuclei bound to the paramagnetic center.

When the τ_s^{-1} values are the dominant terms in Eq. (6), one should substitute τ_{c2} and τ_{c1} for τ_c in Eqs. (7) and (8), the former in the terms with ω_S and the latter in the $4\tau_c$ term and in those with ω_I. The times τ_{c1} and τ_{c2} differ for having T_{1e} and T_{2e}, respectively, instead of τ_s in Eq. (6) (T_{1e} and T_{2e} refer to the electronic relaxation times). However, at low magnetic fields T_{2e} is expected to be equal to T_{1e} and at high magnetic fields the terms containing ω_S tend to zero. Therefore the use of a single τ_s may be justified.

Other equations for T_{1M}^{-1} are available for the case of the splitting of the S manifold (see later).

The contact contribution to T_{1M}^{-1} and T_{2M}^{-1} is given by the following equations [14]:

$$T_{1M(con)}^{-1} = \frac{2}{3}\left(\frac{A}{\hbar}\right)^2 S(S + 1)\frac{\tau_{e1}}{1 + \omega_s^2\tau_{e1}^2} \tag{11}$$

$$T_{2M(con)}^{-1} = \frac{1}{3}\left(\frac{A}{\hbar}\right)^2 S(S + 1)\left(\tau_{e2} + \frac{\tau_{e1}}{1 + \omega_s^2\tau_{e1}^2}\right) \tag{12}$$

where $\tau_{e1,2}^{-1} = T_{1,2e}^{-1} + \tau_M^{-1}$. Here $\tau_{e1,2}$ does not contain τ_r since the contact interaction is not modulated by the molecular tumbling.

In Figure 1 the patterns of the four contributions to nuclear relaxation (T_{1M}^{-1} and T_{2M}^{-1} both contact and dipolar) with increasing magnetic fields are reported for a τ_c of 10^{-11} sec. It appears that at high magnetic fields the dipolar part of T_{1M}^{-1} goes to zero with two steps (dispersions) and the contact part with only one. Also the T_{2M}^{-1} patterns decrease with increasing magnetic field but they do not go to zero because they contain a nondispersive term.

It should be noted that, as far as T_1^{-1} is concerned, the dipolar term is often predominant and anyway its r^{-6} dependence is

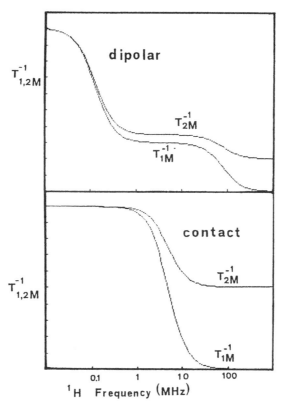

FIG. 1. Pattern of both dipolar and contact contributions to $T_{1,2M}^{-1}$ as a function of the external magnetic field. (Adapted from [1] with permission.)

always qualitatively observed in a set of nonequivalent nuclei. In the case of T_{2M}^{-1} large hyperfine coupling constants and long τ_s (or τ_{c1}) may make the contact term prevailing.

2.3. The Slow-Motion (non-Redfield) Limit

The investigation of nuclear relaxation rates at various magnetic fields is a powerful tool for the understanding of electron-nucleus coupling (see Sec. 3). In many instances, for small molecules τ_c^{-1} is determined by τ_r^{-1}; in other cases, particularly macromolecules,

it is determined by τ_s^{-1}. For high-spin manganese(II) [15,16] and
gadolinium(III) [17], a dependence of τ_s^{-1} on the magnetic field is
found which has been interpreted as being due to the particular
mechanism of electron relaxation which occurs through modulation of
the zero field splitting (ZFS). Even the value of the correlation
time for electron relaxation τ_v is estimated. Therefore a complete
picture comes from the experimental NMR data: $T_{1M}^{-1} < \tau_c^{-1} < \tau_v^{-1}$
(Redfield limit). In the case of high-spin cobalt(II) and nickel(II),
the analysis of the experimental data according to Eq. (7) provides a
τ_c^{-1} about three orders of magnitude larger than in the previous case.
Furthermore, τ_c^{-1} is essentially field-independent and therefore no
information on τ_v can be obtained. There is no experimental evidence
that in the latter cases the overall picture based on the Redfield
limit holds. For this reason a new approach has been developed,
which assumes that molecular tumbling (or other time-dependent
phenomena) modulates the ZFS and thus provides a mechanism for the
electron and the nucleus to relax (slow motion) [18-23]. The system
is still treated in terms of contact and dipolar coupling but now
neither a τ_c for the nucleus-electron interaction nor a τ_v for the
electron-lattice interaction can be defined. The nucleus interacts
with the unpaired electrons and the lattice which together constitute
the generalized lattice [19-23]. The experimental T_{1M}^{-1} of several
systems have been reproduced with this model [19-23].

In the meantime the theory has been developed to include ZFS
and other similar effects within the Redfield limit model [24] (see
Secs. 3.1-3.5). Sample calculations within the two frames have provided the same results under the conditions that the molecular tumbling time in the non-Redfield limit is 2/3 or 4/5 of τ_c in the
Redfield limit for S = 1 or S = 3/2, respectively [25].

These results possibly involve the answer to many questions
regarding the principles of the electron nucleus coupling. For
example, we can say that we can treat a high-spin cobalt(II) system
within the Redfield limit equations even if the system is not in the
Redfield limit. The obtained τ_c may not have a physical meaning but
can still be parametrically used. Alternatively, these results may

be indicative of a very efficient mechanism for electron relaxation which is rotation-independent as well as magnetic field-independent (see Sec. 3.2); in this case the non-Redfield limit would only be an artifact capable of reproducing the experimental results.

2.4. Curie Relaxation

There is a nuclear relaxation pathway which is not bound to electron relaxation but which is still based on the presence of unpaired electrons. In a magnetic field there is a value of $<S_z>$ (see Sec. 2.1) which is independent of the electron relaxation rate. Such value accounts for the presence of a bulk electronic magnetic moment which is measured as magnetic susceptibility. Upon rotation of the molecule, a fluctuating magnetic field is generated at every nucleus; the coupling is again dipolar in nature and its effect on T_{1M}^{-1} and T_{2M}^{-1} is [26,27]:

$$
\begin{aligned}
T_{1M}^{-1} &= \frac{2}{5} \left(\frac{\mu_0}{4\pi}\right)^2 \frac{\gamma_N^2 g_e^2 \mu_B^2}{r^6} <S_z>^2 \frac{3\tau_r}{1 + \omega_I^2 \tau_r^2} \\
&= \frac{2}{5} \left(\frac{\mu_0}{4\pi}\right)^2 \frac{\omega_I^2 g_e^4 \mu_B^4 S^2 (S+1)^2}{(3kT)^2 r^6} \frac{3\tau_r}{1 + \omega_I^2 \tau_r^2}
\end{aligned}
\tag{13}
$$

$$
\begin{aligned}
T_{2M}^{-1} &= \frac{1}{5} \left(\frac{\mu_0}{4\pi}\right)^2 \frac{\gamma_N^2 g_e^2 \mu_B^2}{r^6} <S_z>^2 \left(4\tau_r + \frac{3\tau_r}{1 + \omega_I^2 \tau_r^2}\right) \\
&= \frac{1}{5} \left(\frac{\mu_0}{4\pi}\right)^2 \frac{\omega_I^2 g_e^4 \mu_B^4 S^2 (S+1)^2}{(3kT)^2 r^6} \left(4\tau_r + \frac{3\tau_r}{1 + \omega_I^2 \tau_r^2}\right)
\end{aligned}
\tag{14}
$$

This mechanism is called Curie relaxation because it is related to the bulk susceptibility described by Curie. Since $<S_z>^2$ depends on the square of the external magnetic field, Curie relaxation is effective at high magnetic fields; however, at high magnetic field, T_{1M}^{-1} will level to a small value owing to the term $\omega_I^2/(1 + \omega_I^2 \tau_r^2)$. In

macromolecules, where $\tau_r \gg \tau_s$, the Curie contribution will be smaller
than the dipolar contribution [Eq. (2)] to the overall T_{1M}^{-1}. When T_{2M}^{-1}
is taken into consideration the reverse holds: in macromolecules, the
large τ_r values account for the large line broadening of the signal
at high magnetic field.

3. T_{1M}^{-1} AND NUCLEAR MAGNETIC RELAXATION DISPERSION (NMRD)

NMRD stays for nuclear magnetic relaxation dispersion [28] and is the
study of nuclear relaxation, particularly T_1^{-1}, as a function of magnetic field. Electromagnets can easily be adapted for measurements
down to 4 MHz. There are also specimen of instruments working
between 0.01 and 50 MHz [29]. Many investigations are available
nowadays of T_1^{-1} at various magnetic fields on paramagnetic systems.

We are now going to discuss the fitting of the experimental
profiles of T_{1M}^{-1} vs. the external magnetic field through the calculation of T_{1M}^{-1} within the Redfield limit. Such calculations are often
performed with the Kubo and Tomita formalism [30] starting from a
general Hamiltonian of the type:

$$\hat{H}_0 = \mu_B B_0 g \hat{S} + SDS + \hat{I}A\hat{S} + \hat{S}J\hat{S}_i \tag{15}$$

where D is the ZFS tensor, A is the metal nucleus-unpaired electrons
hyperfine coupling tensor, and J is the magnetic coupling constant
between the S manifold and the S_i manifold of another nearby paramagnetic metal ion. It is assumed that the coupling is dipolar as
indeed it is for protons of coordinated water [31-34] and therefore
an appropriate time-dependent dipolar interaction Hamiltonian is
considered for the computation of T_{1M}^{-1} [35]. A qualitative extension
to the contact contribution is intuitive.

3.1. S = 1/2 Systems

The hexaaqua copper(II) complex shows a water proton T_{1M}^{-1} profile
which can be fitted with Eq. (7) and then with the Hamiltonian (15)

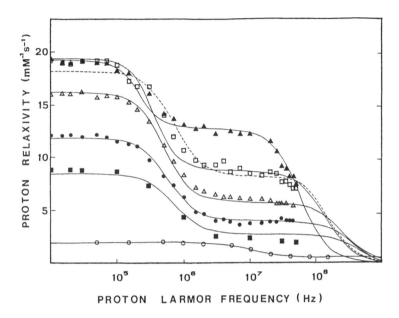

FIG. 2. Solvent proton relaxivities as a function of field for ethylene glycol solutions of $Cu(ClO_4)_2 \cdot 6H_2O$ at -9°C (▲), 5°C (□), 15°C (△), 25°C (●), and 39°C (■) as compared with those of water solutions at 25°C (○). Some best-fit curves obtained according to different approaches are reported [38]. (Reproduced by permission from [38].)

containing only the Zeeman term (Fig. 2) [36]. The τ_c value is 3×10^{-11} sec which corresponds to the rotational correlation time. The G factor, defined as

$$G = \Sigma \left(\frac{n_i}{r_i^6} \right) \tag{16}$$

where i is referred to each proton feeling the paramagnetic center at distance r_i, is 27×10^{-14} pm^{-6}. When the measurements are performed in ethylene glycol, τ_r^{-1} decreases until eventually it becomes smaller than τ_s^{-1} and the latter determines τ_c^{-1} [37,38]. Under these circumstances τ_c^{-1} becomes smaller than $A/\hbar$, i.e., smaller than the copper nucleus-unpaired electron hyperfine coupling. In the magnetic field range where $g\mu_B BS < A/\hbar$, the T_{1M}^{-1} profile is deeply different from that previously seen. Examples are given in Figure 3 and

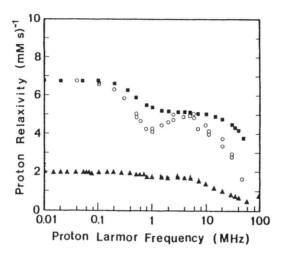

FIG. 3. Water proton NMRD profiles of copper(II) protein solutions: copper(II) bovine carbonic anhydrase (■), copper(II) superoxide dismutase (○), and copper(II) transferrin (▲). (Adapted from [40] with permission.)

TABLE 3

Best Fit Parameters for the Water Proton NMRD Data
on $Cu(II)_2Zn_2$-SOD, $Cu(II)$-BCA II, $Cu(II)_2$-Tf,
and $Cu(II)$-AP

Sample	Temp. (°C)	G^a (pm^{-6})	τ_c (ns)	Ref.
$Cu(II)_2Zn_2SOD$	25	1.7×10^{-15}	1.8	39,40
	15	1.7×10^{-15}	2.5	
	4.5	1.7×10^{-15}	3.8	
	0	1.7×10^{-15}	4.6	
$Cu(II)$-BCA II	25	3.9×10^{-15}	1.9	37,41
$Cu(II)_2$-Tf	25	7.4×10^{-16}	5.7	42
$Cu_2(II)$-AP	25	3.0×10^{-15}	3	43

$^aG = \Sigma(n_i/r_i^6)$, where n_i is the number of equivalent protons interacting with the metal ion at distance r_i. For a single water molecule in the usual coordination geometry, $r = 2.8$ Å and $G = 4.15 \times 10^{-15}$ pm^{-6}.

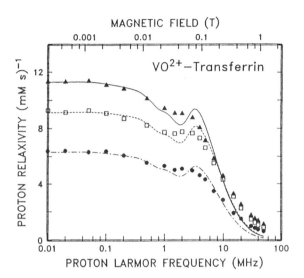

FIG. 4. Water proton NMRD profiles of solutions of VO^{2+}-transferrin at 8°C (▲), 25°C (□), and 38°C (●). (Reproduced by permission from [42].

together with the best fit values and Table 3 shows a list of systems investigated. Also an oxovanadium (IV) derivative of a protein (Fig. 4) has been investigated [42]. Above 4 MHz the Zeeman energy is larger than A/ℏ and Eq. 1 is again suitable for the analysis of the T_{1M}^{-1} profiles of both copper (II) and oxovanadium (IV) systm
systems.

3.2. S = 3/2 Systems

ZFS in high-spin cobalt(II) systems can be as large as 100 cm^{-1} [44-48]. It is common, therefore, that ZFS > $\hbar\tau_c^{-1}$ although τ_c^{-1} can be as large as 10^{12} sec^{-1} (corresponding to 5.3 cm^{-1}). A is never as large as $\hbar\tau_c^{-1}$. Under these circumstances, the Hamiltonian (15) should include the ZFS and Zeeman terms. The effect of ZFS is that of reducing T_{1M}^{-1} and making the dispersion smoother [49]. The T_{1M}^{-1} values are two to four times smaller than expected from Eq. (7) at low field (i.e., when ZFS > Zeeman energy) and G and τ_c values are substantially different in the two analyses (Fig. 5) [50].

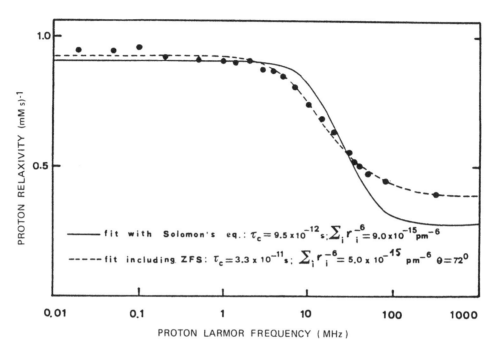

FIG. 5. Water proton NMRD profiles of cobalt(II) carbonic anhydrase at 25°C and pH 9.9. The best-fit curves are calculated according to the Solomon equation (——) or to an equation [Eq. (12) of Ref. 49] taking into account also ZFS (---). The best-fitting parameters resulting from the two approaches are reported [49,50]. (Adapted from [49] with permission.)

The fact that cobalt(II) in water and in ethylene glycol has the same T_{1M}^{-1} profiles indicates that rotation is not involved in the electron and nuclear relaxations [38].

3.3. S = 5/2 Systems

The numerous analyses of high-spin manganese(II) systems are complicated by the magnetic field dependence of τ_c [see Eq. (6)]. Equation (6) together with the field dependence of τ_c [Eq. (17)] does not provide a satisfactory fitting of the experimental data except for the hexaaqua complex [36] since τ_s^{-1} is larger than τ_r^{-1} at low magnetic field but decreases until eventually τ_r^{-1} or τ_M^{-1} may become larger

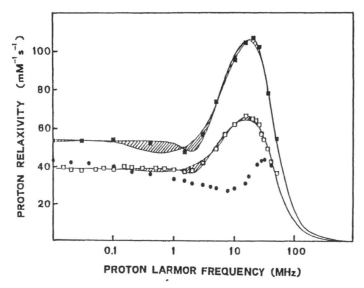

FIG. 6. Water proton NMRD profiles of Mn(II) concanavalin A, at
25°C (□) and 5°C (■), and of Mn(II) carboxypeptidase at 25°C (●)
[53,56]. The solid curves are best-fit curves calculated for D = 0
(upper curve at 0.7 MHz) and D = 0.04 cm^{-1} (lower curve at 0.7 MHz).
(Adapted from [57] with permission.)

[51-53]. Therefore T_{1M}^{-1} is expected to decrease following Eq. (7),
then it starts increasing because τ_s decreases; a maximum is reached
when τ_s^{-1} becomes of the order of τ_r^{-1} or τ_M^{-1} and then a decrease is
observed following the high-field decrease of Eq. (7). Typical pat-
terns are shown in Figure 6 [51-56]. It has recently been shown
that the inclusion of the ZFS is necessary for a quantitative understanding of the experimental pattern [57]. The effect of the manganese nucleus-unpaired electrons hyperfine coupling is minor. The
number of parameters (τ_r, τ_M, τ_s, τ_v, G, ZFS) makes the investigation
of such systems rather difficult.

3.4. S = 7/2 Systems

Gadolinium(III) behaves in a similar way to manganese(II) except that
τ_s is shorter [17]. The aqua complex and the ion in ethylene glycol

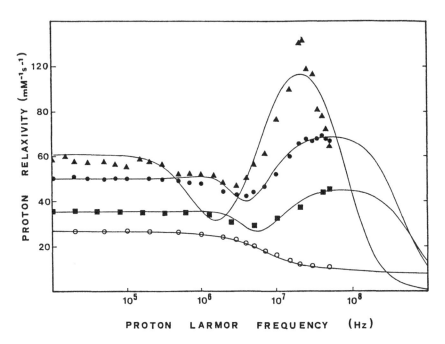

FIG. 7. Solvent proton relaxivities as a function of field for
ethylene glycol solutions of $GdCl_3 \cdot 6H_2O$ at $-9°C$ (▲), $25°C$ (●), and
$39°C$ (■) as compared with those of water solutions at $25°C$ (○). The
calculated best-fit curves are reported [38]. (Reproduced by permission from [38].)

give rise to typical T_{1M}^{-1} patterns (Fig. 7) which have been semi-
quantitatively accounted for by Eq. (7) and field dependence of τ_S
[Eq. (17)] [38]. Again, the inclusion of ZFS would be appropriate.
The pattern of gadolinium(III) in ethylene glycol is similar to that
of Gd^{3+} protein systems [58].

3.5. S = 1 Systems

High-spin nickel(II) is the only non-Kramers' ion for which water
proton T_{1M}^{-1} has been performed as a function of magnetic field and
theoretical treatments are available [18,24,35]. The ZFS accounts
for an almost flat profile at low magnetic fields (Fig. 8) [instead

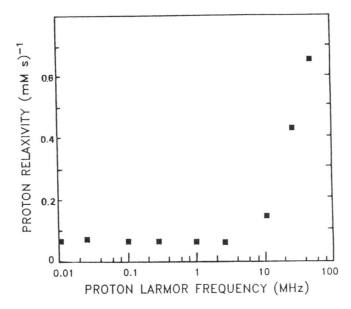

FIG. 8. Water NMRD profile of a solution of Ni(II)-bovine carbonic anhydrase. (Adapted from [24] with permission.)

of the low-field dispersion expected on the basis of Eq. (2)]. It is possible that a field dependence of τ_s induces an increase in T_{1M}^{-1} around 100 MHz in some systems, and that τ_s depends also on τ_r [18,38].

$$4. \quad T_{2M}^{-1}$$

4.1. Factors Contributing to the Linewidth

To our knowledge there is no investigation at low fields of T_{2M}^{-1} or of the linewidth. It is expected, however, that the effects of ZFS and hyperfine coupling will be similar to those discussed for T_{1M}^{-1}. The interest in linewidth studies has been mainly confined in the high-field region. Under these conditions the electronic Zeeman energy is predominant on the other terms of Hamiltonian (15) and therefore they can reasonably be neglected. Indeed, measurements at high magnetic fields of T_1 and T_2 on ^{1}H and ^{2}H have greatly helped in the understanding of the structure of aqua complexes of manganese(II) [53].

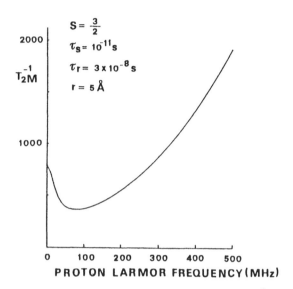

FIG. 9. Field dependence of linewidth of a proton 5 Å distant from a cobalt(II) ion in a system with $\tau_s = 10^{-11}$ sec and $\tau_r = 3 \times 10^{-8}$ sec.

A pattern for T_{2M}^{-1} as a function of the magnetic field is shown in Figure 9. T_{2M}^{-1} decreases with increasing magnetic field following either Eq. (8) or Eq. (12) or both; then it increases again owing to Curie relaxation. The latter contribution depends on B^2 and it is often possible to find a range of magnetic fields, usually above 200 MHz, in which the dipolar or contact interactions have dispersed down or are field-independent. The factorization of Curie relaxation provides a direct tool for obtaining structural information besides the rotational time [59].

4.2. Secondary Relaxation Processes

A nucleus which does not feel the paramagnetic center owing to its distance may be coupled with another nucleus which feels the para-magnetic center. The dipolar coupling between the two nuclei causes an increase of the relaxation rate of the latter nucleus and a decrease of the relaxation rate of the former. There are experi-

mental procedures to detect this nucleus-nucleus interaction, which
is, however, limited to protons owing to their large magnetic moments.
When such an approach is feasible, the extent of the nucleus-nucleus
coupling depends on r^{-6} and can provide conformational information.

5. ELECTRON RELAXATION MECHANISMS

In Table 1 the electron relaxation rates of several metal ions are
reported. Their values depend on the mechanism for the electron
relaxation [60-64]. Let us start with copper(II), S = 1/2. The
direct +1/2 → -1/2 transition is highly improbable at room temperature [65]. Its relative contribution is dominant only at about 1 K.
Another mechanism which is proposed for the solid state is the Raman
process: the scattering of a phonon with the unpaired electron gives
rise to a phonon with less energy and to an electron in the excited
M_S state [65]. Possible phonons are those associated with a vibration of the chromophore; as long as the linewidth of the vibration
is much larger than the ω_s energy for the electron transition, then
this mechanism is possible. Probably analogous mechanisms are operative for solutions as well. The scattering probability depends on
the availability of excited energy levels (coupled to the ground
state via spin orbit): the closer the excited level, the more probable the Raman process [65] (Fig. 10).

   Rotational diffusion mechanisms capable of inducing electron
relaxation [66-74] have been invoked for copper(II) systems; however,
the insensitivity of copper τ_s to rotation makes these mechanisms
less probable. A Raman-like mechanism would also account for the
essential constancy of τ_s in tetragonal copper(II) systems which
display similar electronic spectra.

   On the other extreme unpaired electrons in solid lanthanides
(except Gd^{3+}) have been proposed to relax through an Orbach process
[75]; such process requires two phonons differing by an energy $\hbar\tau_s$
capable of exciting the system to a real excited level (resonant
Raman) (Fig. 11). Such a mechanism is the most efficient in

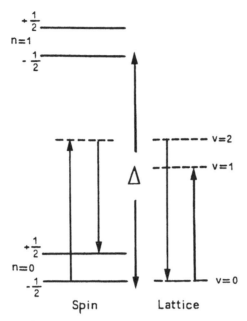

FIG. 10. Schematic representation of a Raman two-phonon process for spin relaxation (see Ref. 1).

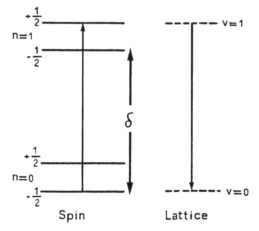

FIG. 11. Schematic representation of an Orbach-type spin relaxation process (see Ref. 1).

performing electron relaxation. Since lanthanides in solution have
the τ_s values of the same order of magnitude as in the solid state
[75,76], a somewhat analogous mechanism may hold. High-spin
cobalt(II), especially if five- or six-coordinated, has very short
τ_s [77]. The presence of low-lying excited states in the latter
stereochemistries would indicate the operativity of Orbach-type
relaxation processes. Possibly this holds also for low-spin iron(III)
(^2T ground state in octahedral symmetry). The impossibility of measuring τ_s in a large range of temperatures in solution prevents us
from being more firm on the actual electronic relaxation mechanism.
The only fact is that the availability of low-lying excited states
makes τ_s shorter. This is, however, consistent with both the Orbach
and Raman processes. It is also consistent with a mechanism which
is called ZFS modulation [11,12]. The ZFS, which occurs only in
S > 1/2 systems, is in general larger the closer the excited states.
Upon rotation, ZFS, which is represented by a traceless tensor, under-
goes modulation. Such modulation is more efficient the larger the
ZFS. In the case of high-spin iron(III), nuclear T_{1M}^{-1} has been found
to be proportional to the square of the ZFS parameter [78]; this has
been taken as evidence that electron relaxation occurs through ZFS
modulation. It is believed that the same mechanism is operative in
high-spin manganese(II). In the case of hexaaqua manganese(II) complex [15,16], instantaneous distortions due to collisions to water
molecules cause a dynamic ZFS which allows the electron to relax.
Electron relaxation is described by the following equation:

$$T_{1e} = \frac{1}{5\tau_{s0}} \left\{ \frac{1}{1 + \omega_s^2\tau_v^2} + \frac{4}{1 + 4\omega_s^2\tau_v^2} \right\} \tag{17}$$

where all the symbols have already been defined. The value of τ_{s0}
is 10^{-10} sec and τ_v is 10^{-12} sec. An equation of this kind is often
used also for manganese(II) and gadolinium(III) proteins. Even in
the case of high-spin six-coordinated nickel(II), which displays
$\tau_s^{-1} = 10^{11}$-10^{10} sec^{-1}, modulation of the ZFS has been suggested as
the main mechanism for electron relaxation [18-23].

It is probable that the latter mechanism may account for
moderate electron relaxation rates. In principle, when electron
relaxation is caused by modulation of the quadratic ZFS in S > 1
systems, there may be more than one electronic relaxation time,
corresponding to allowed transitions between different zero field
split levels [16]. It has been shown that the various relaxation
times coincide, and $T_{1e} = T_{2e}$ as long as $\omega_s\tau_v \ll 1$. In the Redfield
limit, the $\omega_s\tau_v = 1$ dispersion occurs at higher field than the $\omega_s\tau_c$
dispersion, so that when T_{2e} starts to diverge from T_{1e}, nuclear
relaxation is only controlled by the latter (through the ω_I term).
Therefore, the only case where more than one electronic T_1 can be
operative for nuclear relaxation occurs for systems like manganese(II)
or gadolinium(III) in the field region of increase in relaxation
enhancement. Analytical expressions for the two relaxation times
T_{1e}^I and T_{1e}^{II} of S = 3/2 systems have been derived. Only numerical
solutions are available for larger S values.

6.   ELECTRON AND NUCLEUS RELAXATION IN
MAGNETICALLY COUPLED SYSTEMS

6.1.   Effect of Magnetic Coupling on Electron
Relaxation Rates

When a paramagnetic metal ion interacts with a second paramagnetic
metal ion, the electronic relaxation rates of the two metal ions may
be dramatically affected. Let us refer to a metal ion 1, which is
slow relaxing, and to a metal ion 2 which is fast relaxing. The
magnetic coupling is assumed to be isotropic and to be described by
the Hamiltonian $J\underset{\sim}{\hat{S}}_1\cdot\underset{\sim}{\hat{S}}_2$. If J is smaller than $(\hbar\tau_{s1})^{-1}$, the interaction can be taken as negligible. If J is intermediate between
$(\hbar\tau_{s1})^{-1}$ and $(\hbar\tau_{s2})^{-1}$ but still closer to $(\hbar\tau_{s1})^{-1}$, it can be figured out that the electrons of ion 1 will start being relaxed by
coupling with the fast-relaxing electrons of ion 2. The value of
τ_{s1}^{-1} as a function of J can be evaluated perturbatively and expressed
as [79]:

$$\tau_{s1}^{-1}(J) = \tau_{s1}^{-1}(0) + \frac{J^2}{\hbar^2}\tau_{s1}(0) \tag{18}$$

From this equation we can estimate that the τ_s of a copper(II) ion
(3×10^{-9} sec) will go to 10^{-10} sec with a J of 0.1 cm^{-1} upon coupling with a fast-relaxing ion. This case will be further discussed
later. When J is much larger than $(\hbar\tau_{c2})^{-1}$, it can be figured out
that all the relaxation mechanisms operative for the electrons of
ion 2 will be operative also for the electrons of ion 1 and vice
versa. Therefore the system will be described by a single τ_s given
by:

$$\tau_s^{-1} = \tau_{s1}^{-1} + \tau_{s2}^{-1} \tag{19}$$

It is difficult to predict the behavior of the system when J
approaches $(\hbar\tau_{s2})^{-1}$. Of course the above considerations hold as
long as, upon magnetic coupling, no new electronic energy levels
arise which can be efficient for electron relaxation.

In the case of homodinuclear systems, the effect of magnetic
coupling on the electron relaxation rates is small; indeed, the EPR
signals of monomeric and dimeric copper(II) are easily observed
between liquid nitrogen and room temperature. It is conceivable
that in the absence of new mechanisms, under strong magnetic coupling
conditions, the electrons of each ion will relax equally well through
the mechanisms of both ions. Therefore there may be at most a factor
of two in the electronic relaxation rates. At very low temperatures
(4 K) evidence of new pathways for electron relaxation in dicopper(II)
complexes is given [60].

6.2. Solvent ^{1}H T_1^{-1} NMRD of Magnetically Coupled Systems

There is an investigation of glycol ^{1}H T_1 of a solution containing
a bisethylenediamine copper(II) complex interacting with hexacyano-
ferrate(III) (Scheme 1) [79]. The coupling constant J is expected to

be small owing to the orthogonality between the orbitals of copper
and iron which contain the unpaired electrons. An equation has been

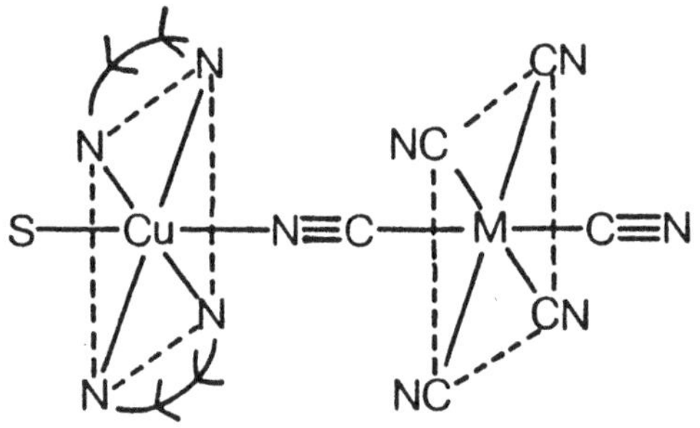

S = SOLVENT

Scheme 1

derived including $J\hat{S}_1 \cdot \hat{S}_2$ in the Hamiltonian (15). However, such
perturbation introduces only a factor 1/2 in nuclear T_{1M}^{-1} values as
long as J << kT. With such an equation a τ_c^{-1} of $\approx 10^{10}$ sec^{-1} has
been obtained for copper(II) which compares with 3×10^8 sec^{-1}
obtained for the bisethylenediamine copper(II) hexacyanocobaltate(III)
system (Fig. 12).

6.3. Nuclear Relaxation Rates in Magnetically
Coupled Systems

The ^{1}H NMR linewidth of the CH_3 signal of the bis(1,2-propanedia-
mine)copper(II) complex decreases with increasing amounts of hexa-
cyanoferrate(III) (Fig. 13) in a way consistent with the τ_s values
given in the previous section [80].

The $[MM'(PMK)_3]^{4+}$ system where PMK = tris-[2,5-di(2-pyridyl)-
3,4-diazahexa-2,4-diene]:

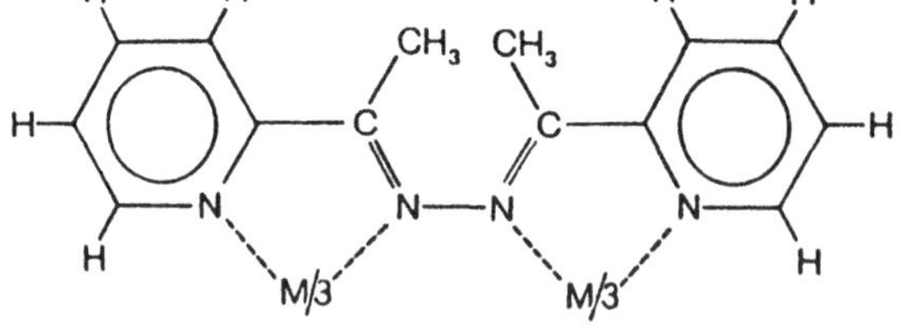

Scheme 2

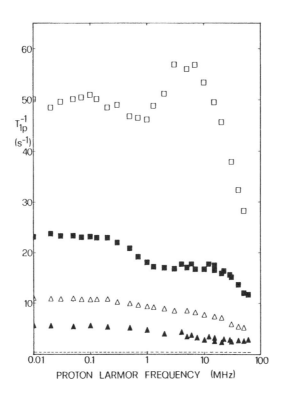

FIG. 12. Solvent NMRD profiles for ethylene glycol solutions of $[Cu(en)_2-Fe(CN)_6]^-$ ($\triangle$ = 0°C, $\blacktriangle$ = 25°C) and $[Cu(en)_2-Co(CN)_6]^-$ ($\square$ = 0°C, $\blacksquare$ = 25°C) in 10 mM concentrations. The dashed line represents the solvent ^1H NMRD profile of $[Fe(CN)_6]^{3-}$ in the same conditions. (Reproduced by permission from [79].)

is a nice system to investigate magnetically coupled systems. As previously anticipated, nuclear T_1^{-1} is proportional to the square of the nucleus-unpaired electron hyperfine coupling. Such hyperfine coupling changes with magnetic coupling; the coefficients to be used are given in Table 4, for all the S_1 and S_2 values encountered in d^n metal ions in the approximation that $J \ll kT$ [81]. At high magnetic fields (360 MHz) the Solomon equation [Eq. (7)] holds for the $[CuZn(PMK)_3]^{4+}$ and $[NiZn(PMK)_3]^{4+}$ complexes. It is found that τ_c is given by τ_r [see Eq. (6)] and is $\cong 1.6 \times 10^{-10}$ sec. In $[Cu_2(PMK_3]^{4+}$, the ^1H T_1^{-1} data are consistent with no coupling at

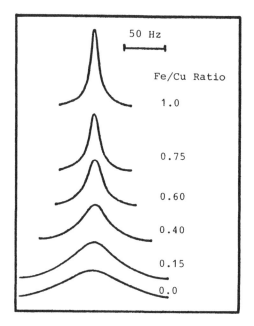

FIG. 13. ^{1}H NMR signal of the methyl group of [Cu(N-propyl-en)$_2$] (ClO$_4$)$_2$ in D$_2$O solution in the presence of K$_3$[Fe(CN)$_6$] at various Fe/Cu ratios. (Reproduced by permission from [80].)

all; indeed, susceptibility measurements show that Curie law is followed down to 5 K [81]. In the case of [Ni$_2$(PMK)$_3$]$^{4+}$ magnetic susceptibility data indicate a J of 15 cm^{-1}. Each proton feels both metal ions to the extents indicated by the Solomon equation multiplied by the factor of Table 4 for S$_1$ = S$_2$ = 1 [81]:

$$T_{1M}^{-1} = K'_{Ni} \left[\frac{1}{r_1^6} + \frac{1}{r_2^6} \right] f(\tau_c) \tag{20}$$

where $f(\tau_c)$ is the term in parentheses in Eq. (7) and K'_{Ni} is the product of the constants in Eq. (7), i.e., K_{Ni} multiplied by the appropriate coefficient of Table 4. The value of τ_c obtained is again τ_r. The effects of magnetic coupling on τ_s is masked by τ_s being larger than τ_r.

In the case of [Co$_2$(PMK)$_3$]$^{4+}$, presumably there is a sizable magnetic coupling [82]; however τ_s^{-1} is relatively large and J has

TABLE 4

X_1 and X_2 Coefficients Calculated for M_1-M_2
Exchange Coupled Pairs [81]

S_1 (columns), S_2 (rows). In each cell the upper value is X_2 and the lower value is X_1.

X_1 \ X_2	5/2	2	3/2	1	1/2
1/2	$\frac{17}{18}$ / $\frac{19}{54}$	$\frac{23}{25}$ / $\frac{9}{25}$	$\frac{7}{8}$ / $\frac{3}{8}$	$\frac{7}{9}$ / $\frac{11}{27}$	1/2 / 1/2
1	$\frac{1049}{1225}$ / $\frac{13}{35}$	$\frac{43}{54}$ / $\frac{7}{18}$	$\frac{467}{675}$ / $\frac{19}{45}$	1/2 / 1/2	
3/2	$\frac{107}{144}$ / $\frac{173}{432}$	$\frac{113}{175}$ / $\frac{379}{875}$	1/2 / 1/2		
2	$\frac{102031}{165375}$ / $\frac{6257}{14175}$	1/2 / 1/2			
5/2	1/2 / 1/2				

(left axis: S_2; top axis: S_1)

to be larger than 20 cm^{-1} in order to be larger than $\hbar\tau_s^{-1}$ and to be effective for electron and nuclear relaxation. Indeed, proton T_{1M}^{-1} are again essentially additive:

$$T_{1M}^{-1} = K_{Co} \left[\frac{1}{r_1^6} + \frac{1}{r_2^6} \right] f(\tau_c) \tag{21}$$

τ_c being the same as in $[CoZn(PMK)_3]^{4+}$ [83]. ^1H T_1^{-1} values for $[CoCu(PMK)_3]^{4+}$ and $[CoNi(PMK)_3]^{4+}$ are given by:

$$T_{1M}^{-1} = X_1 K_{Cu} \left[\frac{1}{r_1^6} \right] f(\tau_{c1}) + X_2 K_{Co} \left[\frac{1}{r_2^6} \right] f(\tau_{c2}) \tag{22}$$

where X_1 and X_2 are the proper coefficients of Table 4. Whereas τ_s of Co^{2+} is that of $[CoZn(PMK)_3]^{4+}$, τ_s of Cu^{2+} in the coupled system is 4×10^{-12} sec and that of Ni^{2+} is 7×10^{-12} sec [81]. A sizable shortening of τ_s has occurred in the slow-relaxing metal ions, although not at such an extent that a single τ_s is displayed by the system. Presumably J is of the order of $\hbar\tau_s^{-1}$.

These studies are of great help in designing strategies for the observation of ^1H NMR in copper(II) proteins. In fact, whereas it is possible to observe the ^1H NMR signals of small copper(II) complexes, especially at high magnetic fields, since τ_c is determined by τ_r ($\simeq 10^{-10}$ sec), it is not possible to observe isotropically shifted ^1H signals for macromolecules for which τ_c is determined by τ_s ($\simeq 10^{-8}$ sec). However, if it is possible to have the copper(II) ion interacting with high-spin cobalt(II), then τ_s of copper can be drastically reduced. This has been found to happen in bovine erythrocyte superoxide dismutase (Cu_2Zn_2SOD); the metal ions are bound as shown in Figure 14 [84].

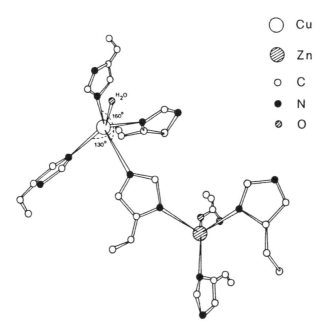

FIG. 14. Coordination spheres of the metal ions in bovine erythrocyte superoxide dismutase (Cu_2Zn_2SOD). (Adapted from [84] with permission.

If zinc is replaced by cobalt(II) there is a magnetic coupling constant of 17 cm^{-1} [85] which is responsible for the decrease of $\tau_s(Cu^{2+})$ toward the value of $\tau_s(Co^{2+})$. The ^{1}H NMR spectrum of Cu_2Co_2SOD is shown in Figure 15 [86]. The signals of protons attached to copper ligands are sharper and have smaller T_{1M}^{-1} than the signals of protons attached to cobalt ligands. The situation is probably similar to $[CoCu(PMK)_3]^{4+}$.

The iron-sulfur proteins are another example in which magnetic coupling reduces the electron relaxation times of the slow-relaxing high-spin iron(III). The ^{1}H NMR signals of oxidized Fe_2S_2 ferredoxins are very broad [87], consistently with two high-spin iron(III) ions. The reduced species contains one high-spin iron(II) (S = 2) which generally relaxes very fast. In this case $J \gg \hbar\tau_s^{-1}$ (90 cm^{-1}) [88] which might lead to a single τ_s^{-1} for the entire system equal to the larger one. The coefficients for nuclear relaxation are not those of Table 4 because $J \cong kT$ and an analysis of the populated levels should be made. The ^{1}H NMR signals are reasonably narrow for all iron-sulfur proteins as shown in Figure 16 [89] for reduced spinach ferredoxin (2Fe2S).

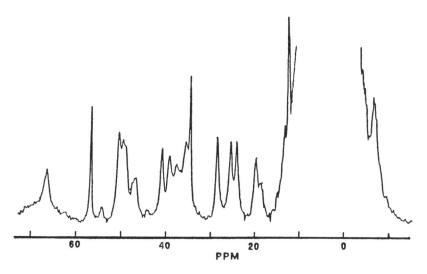

FIG. 15. ^{1}H NMR spectrum (300 MHz) at 30°C of Cu_2Co_2SOD in 10 mM acetate buffer at pH 5.5 in H_2O. (Adapted from [86] with permission.)

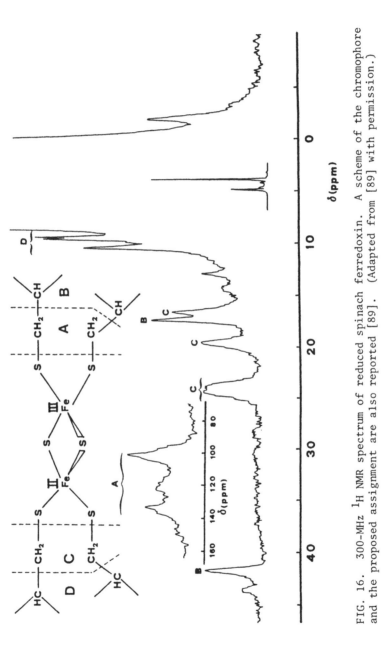

FIG. 16. 300-MHz ^{1}H NMR spectrum of reduced spinach ferredoxin. A scheme of the chromophore and the proposed assignment are also reported [89]. (Adapted from [89] with permission.)

7. CONCLUDING REMARKS

From this survey of the nuclear and electron relaxation properties
it can be learned how to settle the best conditions to further pursue
the NMR research of paramagnetic species in inorganic and bioinorganic
systems. From Table 1 it appears that there are some metal ions which
give reasonably sharp NMR signals [low-spin iron(III), five- and six-
coordinated high-spin cobalt(II), lanthanides(III) except gadolinium-
(III), high-spin iron(II)]. Some others [high-spin manganese(II),
gadolinium(III), and copper(II), especially if they are in macro-
molecules] can be exploited as nuclear relaxing reagents in presence
of excess ligand in exchange with the bulk ligand; otherwise they
would give rise to too broad signals. The possibility of using a
second metal ion of the type of high-spin iron(II) or cobalt(II) to
increase electron relaxation of ions like manganese(II) and copper(II)
should further be exploited.

The choice of the magnetic field is a problem which is often
overlooked. At high magnetic fields, sensitivity and resolution
generally increase; however, in macromolecules, Curie relaxation
leads to severe broadening effects. Therefore, a compromise should
be looked for between the two opposite effects. For example, ^{1}H NMR
spectra of cobalt(II)-substituted ovotransferrin (MW 80,000) and
alkaline phosphatase (MW 90,000) are useless at 300 MHz, whereas they
are sharpest between 60 and 90 MHz. Cu_2Co_2SOD, on the other hand,
can be better studied at 200 or 300 MHz. Curie relaxation depends on
the nuclear magnetic moment and therefore on γ_N. For nuclei other
than ^{1}H, Curie relaxation is negligible for magnetic fields today
accessible. If Curie relaxation is not considered, the high magnetic
field should lead to a sharpening of the line [Eqs. (3) and (5)].
However, there are some suggestions that τ_s in the case of nickel is
field-dependent above 100 MHz. This is probably the reason why in
$[NiZn(PMK)_3]^{4+}$ (Sec. 6.3) τ_c is dominated by τ_r at 360 MHz [79].

Perspectives in the area are a deeper exploitation of nuclear
relaxation even in those cases in which sharp signals are obtained;

unfortunately, this is not often done thus losing quite a share of
information from the NMR experiment. Recently it was shown that
nuclear Overhauser effects can also be observed in paramagnetic
systems [90,91]; if this phenomenon is exploited, a lot of structural information will be gained and a powerful assignment tool will
be available. Also 2D NMR spectroscopy has been reported to be useful in the investigation of polyheme proteins [92].

ABBREVIATIONS

AP	alkaline phosphatase
BCA	bovine carbonic anhydrase
en	ethylenediamine = 1,2-diaminoethane
NMRD	nuclear magnetic relaxation dispersion
SOD	superoxide dismutase
Tf	transferrin
ZFS	zero field splitting

REFERENCES

1. I. Bertini and C. Luchinat, *NMR of Paramagnetic Species in Bio-
    logical Systems,* Benjamin Cummings, Menlo Park, California, 1986.

2. J. Mispelter, M. Momenteau, and J. M. Lhoste, *J. Chem. Soc.
    Dalton,* 1729 (1981).

3. D. M. Doddrell, D. T. Pegg, M. R. Bendall, and A. K. Gregson,
    Chem. Phys. Lett., 40, 142 (1976).

4. H. M. Goff, E. T. Shimomura, and M. A. Phillippi, *Inorg. Chem.,
    22,* 66 (1983).

5. R. M. Golding, R. O. Pascual, and J. Vrbancich, *Mol. Phys., 31,*
    731 (1976).

6. R. J. Kurland and B. R. McGarvey, *J. Magn. Reson., 2,* 282 (1970).

7. G. N. La Mar, W. de W. Horrocks Jr., and R. Holm, *NMR of Para-
    magnetic Molecules,* Academic Press, New York and London, 1973.

8. W. de W. Horrocks, Jr., in *Adv. Inorg. Biochem.,* Vol. 4,
    (G. L. Eichhorn and L. G. Marzilli, eds.), Elsevier, New York,
    1982, p. 201.

9. L. Banci, I. Bertini, and C. Luchinat, in *Rare Earths Spectroscopy* (B. Jezowska Trzebiatowska, J. Legendziewicz, and W. Strek, eds.), World Scientific Publishing, Singapore, 1985.

10. A. G. Redfield, *IBM Research Develop., 1,* 19 (1957).

11. A. G. Redfield, *Adv. Magn. Res., 1,* 1 (1965).

12. R. A. Dwek, *NMR in Biochemistry,* Clarendon Press, Oxford, 1973.

13. I. Solomon, *Phys. Rev., 99,* 449 (1955).

14. N. Bloembergen, *J. Chem. Phys., 27,* 572 (1957).

15. N. Bloembergen and L. O. Morgan, *J. Chem. Phys., 34,* 842 (1961).

16. M. Rubinstein, A. Baram, and Z. Luz, *Mol. Phys., 20,* 67 (1971).

17. S. H. Koenig and M. Epstein, *J. Chem. Phys., 63,* 2279 (1975).

18. H. L. Friedman, M. Holz, H. G. Hertz, and F. Hirata, *J. Chem. Phys., 73,* 6031 (1980).

19. J. Kowalewski, L. Nordenskiöld, N. Benetis, and P. O. Westlund, *Progr. NMR Spectr., 17,* 141 (1985).

20. N. Benetis, J. Kowalewski, L. Nordenskiöld, H. Wennerström, and P. O. Westlund, *Mol. Phys., 48,* 329 (1983).

21. N. Benetis, J. Kowalewski, L. Nordenskiöld, H. Wennerström, and P. O. Westlund, *Mol. Phys., 50,* 515 (1983).

22. N. Benetis, J. Kowalewski, L. Nordenskiöld, H. Wennerström, and P. O. Westlund, *J. Magn. Reson., 58,* 261 (1984).

23. P. O. Westlund, H. Wennerström, L. Nordenskiöld, J. Kowalewski, and N. Benetis, *J. Magn. Reson., 59,* 91 (1984).

24. I. Bertini, C. Luchinat, M. Mancini, and G. Spina, in *Magnetostructural Correlation in Exchange Coupled Systems* (R. D. Willett, D. Gatteschi, and O. Kahn, eds.), Reidel Dordrecht, 1985, p. 421.

25. I. Bertini, C. Luchinat, and J. Kowalewski, *J. Magn. Reson., 62,* 235 (1985).

26. M. Gueron, *J. Magn. Reson., 19,* 58 (1975).

27. A. J. Vega and D. Fiat, *Mol. Phys., 31,* 347 (1976).

28. S. H. Koenig and R. D. Brown, in *ESR and NMR of Paramagnetic Species in Biological and Related Systems* (I. Bertini and R. S. Drago, eds.), Reidel, Dordrecht, 1979.

29. K. Hallenga and S. H. Koenig, *Biochemistry, 15,* 4255 (1976).

30. R. Kubo and K. Tomita, *J. Phys. Soc. Japan, 9,* 888 (1964).

31. D. Waysbort and G. Navon, *J. Chem. Phys., 62,* 1021 (1975).

32. D. Waysbort and G. Navon, *J. Chem. Phys., 68,* 3074 (1978).

33. L. Nordenskiöld, A. Laaksonen, and J. Kowalewski, *J. Am. Chem. Soc., 104,* 379 (1982).

34. J. Kowalewski, A. Laaksonen, L. Nordenskiöld, and M. Blomberg, *J. Chem. Phys.*, *74*, 2927 (1981).

35. U. Lindner, *Ann. Phys. Lpz.*, *16*, 619 (1965).

36. R. Hausser and F. Noack, *Z. Phys.*, *182*, 93 (1964).

37. L. Banci, I. Bertini, and C. Luchinat, *Chem. Phys. Lett.*, *118*, 345 (1985).

38. L. Banci, I. Bertini, and C. Luchinat, *Inorg. Chim. Acta, 100,* 173 (1985).

39. S. H. Koenig and R. D. Brown, III, *Ann. N.Y. Acad. Sci.*, *222,* 752 (1973).

40. I. Bertini, F. Briganti, C. Luchinat, M. Mancini, and G. Spina, *J. Magn. Reson.*, *63,* 41 (1985).

41. I. Bertini and C. Luchinat, in *Biological and Inorganic Copper Chemistry* (K. D. Karlin and J. Zubieta, eds.), Adenine Press, 1986, p. 23.

42. I. Bertini, F. Briganti, C. Luchinat, and S. H. Koenig, *Biochemistry, 24,* 6287 (1985).

43. I. Bertini, C. Luchinat, A. Scozzafava, A. Maldotti, and O. Traverso, *Inorg. Chim. Acta, 78,* 19 (1983).

44. M. B. Yim, L. C. Kuo, and M. W. Makinen, *J. Magn. Reson.*, *46,* 247 (1982).

45. L. C. Kuo and M. W. Makinen, *J. Biol. Chem.*, *257,* 24 (1982).

46. M. W. Makinen and M. B. Yim, *Proc. Natl. Acad. Sci. USA, 78,* 6221 (1981).

47. M. W. Makinen, L. C. Kuo, M. B. Yim, W. Maret, and G. B. Wells, *J. Mol. Cat., 23,* 179 (1984).

48. M. W. Makinen, L. C. Kuo, M. B. Yim, G. B. Wells, J. M. Fukuyama, and J. E. Kim, *J. Am. Chem. Soc.*, *107,* 5245 (1985).

49. I. Bertini, C. Luchinat, M. Mancini, and G. Spina, *J. Magn. Reson.*, *59,* 213 (1984).

50. S. H. Koenig, R. D. Brown, III, I. Bertini, and C. Luchinat, *Biophys. J.*, *41,* 139 (1983).

51. S. H. Koenig, R. D. Brown, III, and J. Studebaker, *Cold Spring Harbor Symp. Quant. Biol.*, *36,* 551 (1971).

52. S. H. Koenig and R. D. Brown, III, in *The Coordination Chemistry of Metalloenzymes* (I. Bertini, R. S. Drago, and C. Luchinat, eds.), Reidel, Dordrecht, 1983, p. 19.

53. T. Kushnir and G. Navon, *J. Magn. Reson.*, *56,* 373 (1984).

54. S. H. Koenig, R. D. Brown, and C. F. Brewer, *Proc. Natl. Acad. Sci. USA, 70,* 475 (1973).

55. S. H. Koenig, R. D. Brown, and C. F. Brewer, *Biochemistry, 24,* 4980 (1985).

56. S. H. Koenig and R. D. Brown, III, *J. Magn. Reson., 61,* 426 (1985).

57. L. Banci, I. Bertini, F. Briganti, and C. Luchinat, *J. Magn. Reson., 66,* 58 (1986).

58. P. B. O'Hara, and S. H. Koenig, *Biochemistry, 25,* 1445 (1986). *Reson. Med. Biol.,* in press.

59. G. N. La Mar, J. T. Jackson, and R. G. Bartsch, *J. Am. Chem. Soc., 103,* 4405 (1981).

60. K. N. Shrivastava, *Phys. Stat. Sol. (b), 117,* 437 (1983).

61. A. A. Manenkov and R. Orbach, *Spin Lattice Relaxation in Ionic Solids,* Harper, New York, 1966.

62. S. A. Al'tshuler and B. M. Kozyrev, *Electron Paramagnetic Resonance* (C. P. Poole, Jr., ed.), Academic Press, New York, 1964.

63. A. Abragam and B. Bleaney, *Electron Paramagnetic Resonance of Transition Ions,* Oxford University Press, London, 1970.

64. H. J. Stapleton, *Magn. Reson. Rev., 1,* 65 (1971).

65. J. H. Van Vleck, *Phys. Rev., 57,* 426 (1940).

66. D. Kivelson, *J. Chem. Phys., 45,* 1324 (1966).

67. H. M. McConnell, *J. Chem. Phys., 25,* 709 (1956).

68. D. Kivelson, *J. Chem. Phys., 33,* 1094 (1960).

69. D. Kivelson, *J. Chem. Phys., 41,* 1904 (1964).

70. R. Wilson and D. Kivelson, *J. Chem. Phys., 44,* 154 (1966).

71. R. Wilson and D. Kivelson, *J. Chem. Phys., 44,* 4440 (1966).

72. P. S. Hubbard, *Phys. Rev., 131,* 1155 (1963).

73. D. Hoel and D. Kivelson, *J. Chem. Phys., 62,* 4535 (1975).

74. W. B. Lewis and L. O. Morgan, *Trans. Met. Chem., Ser. Adv., 4,* 33 (1968).

75. R. Orbach, *Proc. Phys. Soc. (London), A77,* 821 (1961).

76. B. M. Alsaadi, J. C. Rossotti, and R. J. P. Williams, *J. Chem. Soc. Dalton Trans.,* 2147 (1980).

77. I. Bertini, G. Lanini, and C. Luchinat, *J. Am. Chem. Soc., 105,* 5116 (1983).

78. G. N. La Mar and F. A. Walker, *J. Am. Chem. Soc., 95,* 6950 (1973).

79. I. Bertini, G. Lanini, C. Luchinat, M. Mancini, and G. Spina, *J. Magn. Reson., 63,* 56 (1985),

80. I. Bertini, C. Luchinat, F. Mani, and A. Scozzafava, *Inorg. Chem., 19,* 1333 (1980).

81. C. Owens, R. S. Drago, I. Bertini, C. Luchinat, and L. Banci, *J. Am. Chem. Soc.,* in press.

82. C. Benelli, A. Dei, and D. Gatteschi, *Inorg. Chem., 21,* 1284 (1982).

83. I. Bertini, C. Luchinat, C. Owens, and R. S. Drago, in preparation.

84. J. A. Tainer, E. D. Getzoff, K. M. Beem, J. S. Richardson, and D. C. Richardson, *J. Mol. Biol., 160,* 181 (1982).

85. I. Morgenstern-Badarau, D. Cocco, A. Desideri, G. Rotilio, J. Jordanov, and N. Dupré, *J. Am. Chem. Soc., 108,* 300 (1986).

86. I. Bertini, G. Lanini, C. Luchinat, L. Messori, R. Monnanni, and A. Scozzafava, *J. Am. Chem. Soc., 107,* 4391 (1985).

87. R. H. Holm and J. A. Ibers, in *Iron Sulphur Proteins* (W. Lowenberg, ed.), Chap. 7, Academic Press, New York, 1977.

88. J. P. Gayda, P. Bertrand, C. More, and R. C. Cammack, *Biochimie, 63,* 847 (1981).

89. I. Bertini, G. Lanini, and C. Luchinat, *Inorg. Chem., 23,* 2729 (1984).

90. S. Ramaprasad, R. D. Johnson, and G. N. La Mar, *J. Am. Chem. Soc., 106,* 3632 (1984).

91. S. Ramaprasad, R. D. Johnson, and G. N. La Mar, *J. Am. Chem. Soc., 106,* 5330 (1984).

92. H. Santos, D. L. Turner, A. V. Xavier, and J. Le Gall, *J. Magn. Reson., 59,* 177 (1984).

3

NMR Studies of Magnetically Coupled Metalloproteins

Lawrence Que, Jr.
Department of Chemistry
University of Minnesota
Minneapolis, Minnesota 55455

and

Michael J. Maroney
Department of Chemistry
University of Massachusetts
Amherst, Massachusetts 01003

1. INTRODUCTION

In recent years, the number of metalloproteins recognized as containing magnetically coupled binuclear active sites has increased dramatically and NMR studies have been useful in providing insights into the properties of these metal clusters. These studies have taken advantage of the unique properties derived from the antiferromagnetic interaction between metal centers to probe the structure, function, and magnetic properties of the active sites. To be covered in this chapter are studies on the Cu(II)Co(II) derivative of bovine superoxide dismutase, the various oxidation states of hemerythrin, ribonucleotide reductase, uteroferrin, and two-iron ferredoxins.

2. PHYSICAL BACKGROUND

2.1. The Paramagnetic Shift

The presence of a paramagnetic center in a complex can induce shifts in the ligand proton resonances such that they lie outside the region expected for diamagnetic compounds [1]. Such shifts arise via either a through-bond (contact) or a through-space (dipolar) interaction of the nucleus with the paramagnetic center [2]. The contact interaction is proportional to χ (the magnetic susceptibility of the complex) and A (the electron nuclear hyperfine splitting constant). A is a measure of the delocalization of unpaired spin density from the paramagnetic center to the nucleus in question via the σ or π bonding framework. Shifts arising from a σ-delocalization mechanism are characterized by a rapid attenuation of the paramagnetic effect. Such shifts are always downfield of the diamagnetic position, and a decrease of an order of magnitude with each intervening bond is typical. On the other hand, shifts derived from a π-delocalization mechanism do not attenuate with each intervening bond and are characterized by alternating in signs, thus allowing the possibility of upfield shifts.

The dipolar interaction depends on the presence of a magnetically anisotropic paramagnetic center. In an axially anisotropic system, i.e., $g_\parallel \neq g_\perp$, the dipolar effect on the shift depends on the position of the nucleus in question relative to the paramagnet. This relationship is described in Eq. (1):

$$\delta_{\text{dipolar}} \propto (\chi_\parallel - \chi_\perp) \frac{3 \cos^2\theta - 1}{r^3} \tag{1}$$

where θ is the angle made by the metal-nucleus vector relative to the z axis of the magnetic anisotropy and r is the metal-proton distance. The dipolar shift can be positive or negative and is zero at $\theta = 54.7°$. For our discussion, high-spin Fe(III) has a spherically symmetric ground electronic state and thus gives rise to negligible dipolar contributions. However, Cu(II), Co(II), and Fe(II) have electronic configurations that can give rise to significant magnetic anisotropy.

2.2. The Paramagnetic Linewidth

The presence of the paramagnetic center dramatically increases the linewidths [3]. In all cases to be discussed, the dominant factor determining linewidth is the electronic spin-lattice relaxation time. The shorter T_{1e} becomes, the sharper the signals obtained.

In this discussion, the metals involved are Cu(II), Co(II), Fe(III), and Fe(II), all in their high-spin configurations. Of these, T_{1e} for Cu(II) is the longest (10^{-7} sec), its EPR signal being observable at room temperature. The NMR spectra of Cu(II) complexes are normally unobservable due to the broadness of the resonances. High-spin Fe(III) has the next longest T_{1e} (10^{-8}-10^{-11} sec), depending on the zero field splitting, which is affected by the ligand environment [4]. The larger the zero field splitting, the more efficient the electronic relaxation. Co(II) and Fe(II) have T_{1e}'s of approximately 10^{-12} sec and give rise to the sharpest features.

2.3. Antiferromagnetic Coupling

The interaction of two paramagnetic metal centers in a binuclear complex can result in the antiferromagnetic coupling of the two centers. The interaction, which can be mediated by a bridging ligand atom, can be described with the spin Hamiltonian [Eq. (2)]:

$$\hat{H} = -2JS_1 \cdot S_2 \tag{2}$$

where S_1 and S_2 are the spin vectors for the two centers and J is the magnitude of the interaction. For a system where $S_1 \geq S_2$, antiferromagnetic coupling would lead to a manifold of spin states $(S' = S_1 - S_2, S_1 - S_2 + 1, \ldots, S_1 + S_2)$ at energies of $J[S'(S' + 1)]$ [5]. The population of these states as a function of temperature would depend on the value of J. From 0 K, χ would increase with T, maximize at the Néel temperature, and then decrease with increasing T. The contact shift observed, being proportional to the magnetic susceptibility, is a reflection of the population of the various spin states and provides an estimate of the value for J. Such considerations will be useful for interpreting the spectra of the binuclear proteins.

The other major effect of antiferromagnetic coupling is the linewidth. When a slow-relaxing metal center interacts with a faster relaxing metal center, the cluster adopts the T_{1e} of the faster relaxing center [Eq. (3)].

$$\frac{1}{T_{1e}}(M - M') = \frac{1}{T_{1e}}(M) + \frac{1}{T_{1e}}(M') \tag{3}$$

Thus, the resulting NMR spectrum of a binuclear complex would be significantly sharper than one derived from a complex of the slow-relaxing metal center alone [6]. The magnetic interaction required need not be large [7] to achieve dramatic improvements in spectral quality. The significant narrowing of linewidths as a result of this has facilitated the observation of the NMR spectra of Co_2Cu_2-superoxide dismutase and the mixed-valent forms of hemerythrin and uteroferrin.

3. SUPEROXIDE DISMUTASE

Bovine erythrocyte superoxide dismutase (SOD) contains a hetero-binuclear Zn(II)-Cu(II) cluster that catalyzes the disproportiona-tion of superoxide to dioxygen and hydrogen peroxide [8,9]. The enzyme mechanism is believed to involve coordination of superoxide to Cu and subsequent redox chemistry at the Cu site [8]. The metal ions in the active site are bridged by an imidazolate group (His 61). The zinc coordination sphere is completed by two other imidazoles (His 69 and 78) and a carboxylate (Asp 81). The copper is further coordinated by three imidazoles (His 44, 46, and 118) in a flattened tetrahedron and an apical water molecule.

The native enzyme exhibits a rather rhombic EPR spectrum arising from the Cu(II) center [7]; the addition of anions such as halides, CN^-, N_3^-, NCO^-, and NCS^- results in a shift to more axial symmetry [10]. The Zn(II) center can be replaced by a Co(II) ion to afford an enzyme of comparable activity. The typical Cu(II) EPR signal of the native enzyme is not observed in the Co_2Cu_2 enzyme due to the antiferromagnetic coupling of the two metal centers [11]. Upon addition of N_3^-, this EPR signal does not reappear, indicating that the imidazolate bridge is retained [12].

Bertini and coworkers have taken advantage of the shorter T_{1e} of Co(II) centers to probe the active site of SOD with NMR spectros-copy [13,14]. Their results provide insight into the coordination chemistry of the binuclear unit upon anion binding and upon reduc-tion of the Cu(II) center, both processes being features of the proposed catalytic mechanism.

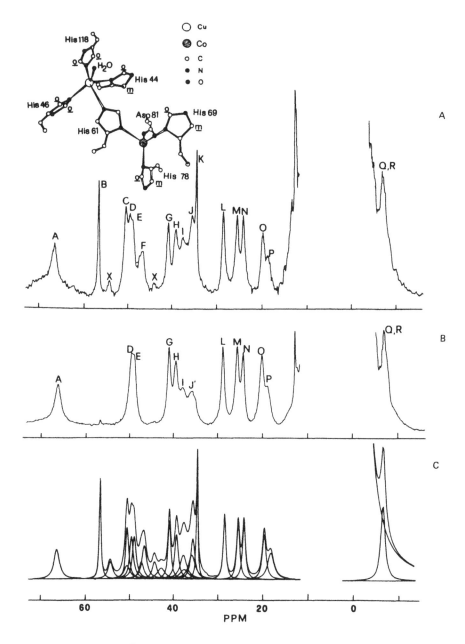

FIG. 1. 300-MHz ^{1}H NMR spectrum at 30°C of $Cu_2^{II}Co_2^{II}$SOD in 10 mM
acetate buffer pH 5.5 in H_2O (a) and in D_2O (b). (c) Computer simulation of (a). (Reprinted with permission from *J. Am. Chem. Soc.*,
107, 4391 (1985) [13]. Copyright 1985 American Chemical Society.)

The NMR spectra of $Cu(II)_2Co(II)_2$ SOD in H_2O and D_2O are shown in Figure 1 and can be compared with those of $Cu(I)_2Co(II)_2SOD$ in Figure 2 [13,14]. Features arising from Co(II) ligands may be identified from the latter spectrum since Cu(I) is diamagnetic. These features strongly resemble those observed for the $E_2Co(II)_2SOD$ derivative where the copper site is vacant. The resonances of $Cu(II)_2Co(II)_2SOD$ exhibit linewidths which are similar to or smaller than the corresponding features in either of the other two derivatives. This indicates that the magnetic interaction between the two metal centers is large enough to result in a cluster T_{1e} that is equal to or shorter than that for Co(II). The additional resonances observed in the $Cu(II)_2Co(II)_2SOD$ spectrum arise from the ligands bound to the Cu(II) site, which in the absence of magnetic coupling between Cu(II) and Co(II) would be too broad to be observed because of the large T_{1e} characteristic of Cu(II). The Cu-ligand resonances

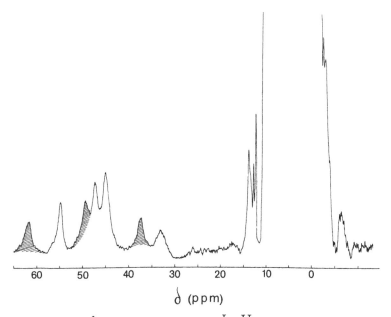

FIG. 2. 300-MHz ^1H NMR spectrum of $Cu_2^ICo_2^{II}SOD$ at pH 5.5 at 30°C in H_2O. Shaded features represent solvent exchangeable resonances. (Adapted with permission from *J. Am. Chem. Soc.*, *107*, 2178 (1985) [14]. Copyright 1985 American Chemical Society.)

are in fact sharper than those for the Co(II)-bound ligands because
proton dipolar relaxation [3] is a function of the magnetic susceptibility of the metal center with which it is associated, given a
single cluster T_{1e}. Thus the ligand protons on Cu(II) reflect the
effects of an $S = 1/2$ center, while those on Co(II) reflect the
effects of an $S = 3/2$ center.

The assignment of the spectrum of $Cu(II)_2Co(II)_2SOD$ (Table 1)
is based on a variety of observations including the effects on the
NMR spectra of H_2O/D_2O solvent exchange, proton T_1 measurements, and
the known x-ray crystal structure of the native $Cu(II)_2Zn(II)_2SOD$.
The structure [9] shows that the three histidines in the Zn(Co) site
are coordinated via the δ nitrogen, while only one of the four histidines in the Cu site is coordinated via the δ nitrogen, the other
three being bound via the ε nitrogen (Figure 1). The imidazole N-H's
are easily assigned because they exchange with solvent. Three of the
five imidazole N-H's immediately disappear in D_2O (B, J, K); these
are assigned to the histidines coordinated to Cu(II). The other two
imidazole N-H's can be exchanged by treating $E_2Co(II)_2SOD$ with D_2O
and then adding Cu(II) to form $Cu(II)_2Co(II)_2SOD$. These are assigned
(C, F) to the histidines coordinated to Co(II). The slower rate of
exchange is consistent with the inaccessibility of the Co site to
solvent as indicated by the x-ray structure. The signals labeled
C and F are also noticeably broader than those labeled B, J, and K,
consistent with their assignment to the more paramagnetic metal
center.

Table 1 lists the assignments made by Bertini et al. [13]
based on their observations. Note that the N-H protons on imidazole
coordinated to the Cu(II) center are the sharpest and exhibit the
longest T_1's. This is due to their large distance from the metal
center. Similarly, the meta H on His 44 [the only His bound to
Cu(II) via the δ nitrogen] has T_1 and T_2 values like the N-H on
His 44 because of the similar distance from the Cu(II) center. The
meta-H protons on histidines coordinated to Co(II) have somewhat
shorter T_1's because of the higher paramagnetism of the Co(II).

TABLE 1

NMR Parameters of the Signals of
Cu_2Co_2SOD at 300 MHz and 30°C

Signal[a]	Chem shift (ppm from TMS)	Linewidth[b] (Hz)	T_1 (msec)	Proposed assignment
A	66.2	430	1.1	o-H(Co), o-H(Cu) (His 61)
B	56.5	115	4.1	NH(Cu)
C	50.3	240	-[c]	NH(Cu)
D	49.4	295	3.1	m-H(Co)
E	48.8	295	3.1	m-H(Co)
F	46.7	380	-[c]	NH(Co)
G	40.6	210	2.8	o-H(Cu)
H	39.0	280	1.7	o-H(Cu)
I	37.4	530	1.4	o-H(Co)
J'	35.6	530	1.6	o-H(Co)
J	35.4	310	-[c]	NH(Co)
K	34.5	105	4.5	NH(Cu)
L	28.4	190	4.2	m-H(Cu)
M	25.3	220	2.5	o-H(Cu)
N	24.1	220	2.5	o-H(Cu)
O	19.6	280	2.4	o-H(Cu)
P	18.7	480	1.2	o-H(Cu), m-H(Co) (His 61)
Q	-6.2	330	2.2	β-CH_2
R	-6.2	330	2.2	β-CH_2

[a]The signals are labeled according to Figure 1.
[b]Obtained from the spectral simulation in Figure 1.
[c]Not measured because the signal is within a complex envelope.
Source: Reprinted with permission from [13]. Copyright 1985 by
American Chemical Society.

The NMR spectrum has been used to probe changes in the active site. The addition of anions like N_3^- and NCO^- causes changes in the isotropic shifts of many resonances [13]. In particular, peaks K, L, and O move dramatically toward their diamagnetic positions. Peak O is shifted into the diamagnetic region while peaks K and L shift to 17 and 14 ppm, respectively, in the N_3^- adduct. These are peaks assigned to His 44 and their shifts toward the diamagnetic region indicate that anion binding results in its displacement from the Cu(II) site. The small isotropic shift still remaining is probably due to dipolar effects from the Co(II) center. The other resonances associated with the Cu(II) site are also significantly perturbed by the binding of anion. This indicates some structural reorganization in the active site [13], which is also reflected by changes in EPR spectra upon binding [11].

Another study reported by Bertini et al. [14] involves the active site structure of $Cu(I)_2Co(II)_2SOD$, where only Co(II)-ligand protons are observed outside the diamagnetic region of the spectrum. Three NMR features are lost upon solvent deuteration, one more than would be expected on the basis of the $Cu(II)_2Zn(II)_2SOD$ structure. This demonstrates that the imidazolate bridge between the metals is broken in the $Cu(I)_2Co(II)_2$ form, the additional exchangeable proton being a substitution for the Cu(I) center on the bridging imidazole. Such a picture is consistent with previously reported observations. For example, (1) the uptake of a proton upon reduction of Cu(II) to Cu(I) in native SOD [15], (2) the similarity of the Co(II) electronic spectra in $Cu(I)_2Co(II)_2SOD$ and $E_2Co(II)_2SOD$ [16], (3) the near identity of ^{113}Cd NMR chemical shifts in $Cu(I)_2Cd(II)_2SOD$ and $E_2Cd(II)_2SOD$ [17], and (4) EXAFS observations that Cu(I) in $Cu(I)_2Zn(II)_2SOD$ is three-coordinate [18].

Thus, NMR studies of Cu_2Co_2SOD have been useful for obtaining insights into the structure of the active site under conditions distinct from those employed in the x-ray structure determination and for examining structural changes that may occur along the reaction path, namely, superoxide binding and reduction of the Cu center.

4. HEMERYTHRIN

Hemerythrin (Hr), a dioxygen carrier found in a number of marine invertebrates, has a binuclear iron unit as its active site [19-21]. The redox process that takes place during the oxygen-binding reaction couples the oxidation of two high-spin ferrous centers to high-spin ferric centers with the reduction of oxygen to peroxide. Mössbauer spectroscopy [22-24] clearly demonstrates the oxidation state changes of the iron centers and resonance Raman spectra [25,26] reveal an O-O stretch characteristic of peroxide in oxyhemerythrin. X-ray crystallographic studies [21] show that in $[Fe(III)]_2$ forms of the protein the iron atoms are bridged by an oxo group and two endogenous carboxylates (Asp 106, Glu 58). The outer faces of the confacial bioctahedron are capped by three histidines on one side and two histidines and an exogenous ligand on the other side. This exogenous ligand is hydroperoxide in oxyHr, azide in $metHrN_3$, and absent in metHr.

Three oxidation states of this protein have been characterized spectroscopically: $[Fe(II)]_2$ (deoxy), [Fe(III), Fe(II)] (semimet), and [Fe(III), Fe(III)] (oxy and met). NMR studies have been useful for providing some insights into the coordination chemistry of the binuclear cluster.

4.1. Met Complexes

The paramagnetically shifted regions of the NMR spectra of *Phascolopsis gouldii* hemerythrin in various met forms are shown in Figure 3. Exchangeable resonances are observed in the 17 to 23 ppm region, while a nonexchangeable feature is found at 11 ppm [27]. The exchangeable signals are readily assigned to His N-H's coordinated to the binuclear cluster. Their isotropic shifts (Table 2) can be compared with those of imidazoles coordinated to mononuclear high-spin Fe(III) centers (Table 3). The significant differences in the shifts can be attributed to the presence of antiferromagnetic coupling between the iron centers in hemerythrin. Magnetic susceptibility measurements

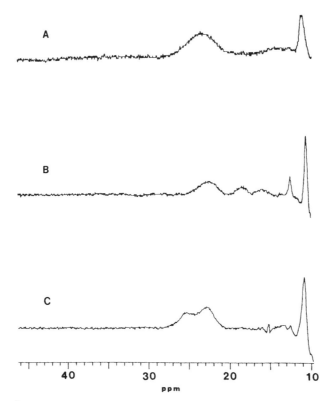

FIG. 3. ^1H NMR spectra of methemerythrin complexes in 50 mM phosphate buffer, pH 7.5, at 30°C.   (A) metHr, (B) metHrN$_3$, (C) metHrS.
(Reprinted with permission from *J. Am. Chem. Soc., 106,* 6445 (1984)
[27].   Copyright 1984 American Chemical Society.)

show that J = -77 cm^{-1} for oxyHr and -134 cm^{-1} for metHr [40].   At
room temperature, this would correspond to an approximately 90%
decrease in χ relative to the uncoupled situation and results in a
reduced contact shift.

The nonexchangeable feature at 11 ppm is assigned to the CH$_2$
protons adjacent to the bridging carboxylates [27].   This assignment
is based on the observation of acetate CH$_3$ signals in hemerythrin
model compounds in the same region (Table 3).   Armstrong and Lippard
[38] assign the CH$_3$ resonance in the spectrum of [(HBpz$_3$Fe)$_2$O(OAc)$_2$]
to a signal of 10.5 ppm and we have made a similar assignment in the

TABLE 2

Paramagnetically Shifted Features in the NMR Spectra of Binuclear Iron Proteins

Complex	T (°C)	Chemical shifts (ppm)	Ref.
metHrN$_3$	30	23, 19, 16, 13 (His NH); 11 (Asp and/or Glu CH$_2$)	27
metHr	30	24, 13 (His NH); 11 (Asp and/or Glu CH$_2$)	27
metHrS	30	25, 23 (His NH); 11 (Asp and/or Glu CH$_2$)	27
semimetHrN$_3$	40	72.2 (His NH, Fe(III)); 53.6 (His NH, Fe(II)); 20, 15, 13, -1.1 (unassigned)	27
semimetHrS	40	54.0, 43.6, 37.2, 32.1, 22.5 (His NH); 16, 15, 12, -1.2, -6.1 (unassigned)	27
deoxyHr	40	62.4, 46.3, 43.2 (His NH); 14, 11, -1.9, -2.5, -3.1, -5.8 (unassigned)	28
deoxyHrN$_3$	40	78.3, 66.8, 47.3 (His NH); 25, 20, 19, 16, 15, 14, -1.9, -3.3, -10 (unassigned)	28
Ribonucleotide reductase	10	24 (His NH); 19 (unassigned)	29
Uteroferrin	40	87 (His NH, Fe(III)); 87 (unassigned, Fe(II)); 69, 61 (Tyr 2,6-H, Fe(III)); 43 (Tyr β-CH$_2$, Fe(III)); 43 (His NH, Fe(II)); 29, 23, -24 (unassigned); -69 (Tyr 3,5-H, Fe(III))	30
Uteroferrin-MoO$_4$ complex	40	102 (unassigned, Fe(II)); 85 (His NH, Fe(III)); 66 (Tyr 2, 6-H, Fe(III)); 51 (His NH, Fe(II)); 43 (Tyr β-CH$_2$, Fe(III)); 25, -5, -8, -21 (unassigned); -70 (Tyr 3,5-H, Fe(III))	a
Uteroferrin-WO$_4$ complex	40	110 (unassigned, Fe(II)); 87 (His NH, Fe(III)); 71, (Tyr 2, 6-H, Fe(III)); 54 (His NH, Fe(II)); 46, 44 (Tyr β-CH$_2$, Fe(III)); 28, 25, -5, -11, -23 (unassigned); -74 (Tyr 3,5-H, Fe(III))	a

[a]J. W. Pyrz and L. Que, Jr., unpublished observations.

TABLE 3

Shifts of Ligand Protons in Iron Complexes

Complex	Shifts	Ref.
Phenolates		
Fe(III)		
Fe(salen)(OC$_6$H$_4$-4-CH$_3$)	-101(o-H); 89(m-H); 110(p-CH$_3$)	31
Fe(salen)(TyrOMe)	-96(3',5'-H); 87(2',6'-H); 53, 73(β-CH$_2$)	30
Fe(salen)(NAcTyrOMe)	-97(3',5'-H); 85(2',6'-H); 58, 61(β-CH$_2$)	30
Fe(salhis)$_2^+$	-79(sal 3-H); 76(sal 4-H), -66(sal 5-H); 48(sal 6-H)	30
Fe(II)		
Fe(salhis)$_2$	-18(sal 3-H); 41(sal 4-H); -5(sal 5-H); 31(sal 6-H)	30
[MoS$_4$Fe(OC$_6$H$_4$-4-CH$_3$)$_2$]$^{2-}$	-43(o-H); 41(m-H); 46(p-CH$_3$)	32
Imidazoles		
Fe(III)		
Fe(salhis)$_2^+$	104(N-H); 81(im 4-H)	30
[Fe(salen)im$_2$]$^+$	103(N-H); 78(im 4-H)	30
[Fe(salen)(4-Meim)$_2$]$^+$	103(N-H); 19(4-CH$_3$)	30
[Fe(salpyr)$_2$(4-Meim)$_2$]$^+$	97(N-H); 20(4-CH$_3$)	30
Fe(II)		
Fe(salhis)$_2$	57(N-H)	30
Fe(biim)$_3^{2+}$	68(N-H)	30
Fe-bleomycin A$_2$	67(N-H)	33
Fe(N-MeTPP)im	65(N-H)	34
Fe[T(p-iso-PrP)P]2Meim	73(N-H)	35
Carboxylates		
Fe(III)		
Fe(salen)OAc	144(CH$_3$)	36
[(HBpz$_3$Fe)$_2$O(OAc)$_2$] (J = -121 cm^{-1})	10.5(CH$_3$)	37
[(HBpz$_3$Fe)$_2$OH(OAc)$_2$]$^+$ (J = -17 cm^{-1})	69(CH$_3$)	38
[(Me$_3$tacnFe)$_2$O(OAc)$_2$]$^{2+}$ (J ~ -100 cm^{-1})	11.0(CH$_3$)[a]	39

[a]M. J. Maroney, K. Wieghardt, and L. Que, Jr., unpublished results.

spectrum of $[(tacnFe)_2O(OAc)_2]^{2+}$. Thus the NMR spectra of these Hr complexes in solution are consistent with their solid state structures.

4.2. Semimet Forms

Mixed valent forms of hemerythrin can be made by one-electron reduction of metHr or one-electron oxidation of deoxyHr [41]. The EPR spectra of the resulting complexes are different, but the addition of N_3^- to either complex yields a single semimetHrN$_3$ complex. These complexes have EPR g values all <2, typical of an antiferromagnetically coupled Fe(III)-Fe(II) pair. The EPR signals arise from the ground S = 1/2 state, which has a T_{1e} short enough that the signals are unobservable above 40 K.

The semimetHrN$_3$ complex is reasonably stable near room temperature and exhibits an NMR spectrum shown in Figure 4 [27]. Two solvent exchangeable resonances are found at 72 and 54 ppm at 40°C and assigned to N-H's on histidines coordinated to the Fe(III) and Fe(II) sites, respectively. These assignments are based on the premise that the metal to which the ligand is coordinated makes the dominant contribution to the isotropic shifts of the ligand protons. This has been demonstrated by Gatteschi et al. [6] in weakly coupled heterobinuclear complexes and the assignments are consistent with the observed shifts for NH protons on imidazole ligands coordinated to high-spin ferric and ferrous centers (Table 3) after correction for the antiferromagnetic interaction. Furthermore, the ratio of the areas of the 54 ppm peak and the 73 ppm peak is 3:2, demonstrating that the iron atom bound to three histidines is the ferrous center. The azide ligand would be expected to bind preferentially to the ferric center and the persistence of the azide-to-Fe(III) charge transfer band in semimetHrN$_3$ [41] is consistent with the assignments.

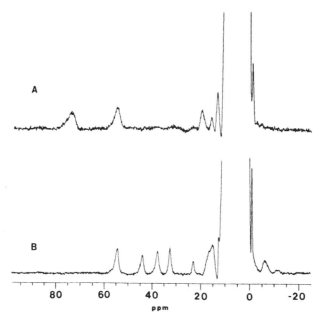

FIG. 4. 300-MHz ^{1}H-NMR spectra of semimethemerythrins in 50 mM
phosphate buffer, pH 7.5. (A) semimetHrN$_3$ at 35°C. (B) semimetHrS
at 40°C. (Reprinted with permission from *J. Am. Chem. Soc., 106,*
6445 (1984) [27]. Copyright 1984 American Chemical Society.)

        The magnitude of the antiferromagnetic coupling J can also be
obtained from the NMR data. Because the contact shift is proportional to χ, the temperature dependence of the shift can in principle
be fit to the temperature dependence of χ in an antiferromagnetic
system to extract J. Two things need to be considered. First, this
approach assumes that the A value of a particular nucleus is invariant
with spin state (S') and that the dipolar component is small. Studies
on three model systems indicate the limitations to this approach. The
three complexes are $[Fe_2S_2(S_2\text{-o-xyl})_2]^{2-}$ (J = -148 cm^{-1}) [42],
$[Fe(salen)]_2O$ (J = -90 cm^{-1}) [36], and $[Fe_2(sal_3trien)(OMe)Cl_2]$
(J = -8 cm^{-1}) [30,43]. All are nonheme iron(III) complexes with
small zero field splittings, so the dipolar contribution to the
isotropic shift would be small. The first and third complexes have
isotropic shift temperature dependences that match well with those

of their respective χ's, while [Fe(salen)]$_2$O shows some significant discrepancies.

The deviations in the [Fe(salen)]$_2$O case have been attributed to the spin state variation of A [36]. For the other two cases, the effect of the variable A is minimized because J is quite large in one case and rather small in the other with respect to kT. The coupling in [Fe$_2$S$_2$(S$_2$-o-xyl)$_2$]$^{2-}$ is large enough so that only the S' = 0 and S' = 1 states need be considered in the temperature range studied; only the latter is paramagnetic and thus only one A value need be considered. On the other hand, the coupling in [Fe$_2$(sal$_3$-trien)(OMe)Cl$_2$] is so small that all spin states are populated at room temperature, and the variations in shift due to changes in the population of the S' = 5 state become less pronounced because of the population of other paramagnetic states. In a similar study on [(HBpz$_3$Fe)$_2$OH(OAc)$_2$]$^+$ (J = -17 cm^{-1}), we found agreement between the temperature dependences of the shift and χ [44]. In cases where J is small, χ is quite sensitive to changes in J so that a reasonable estimate of J can be obtained despite some dependence of the A value on spin state.

The second aspect to consider is how to relate χ of the cluster to χ of the individual component metal center. In the examples above, both metal centers are the same, high-spin Fe(III), so that the contribution of each iron center to the cluster susceptibility is half the measured cluster susceptibility. In the semimetHr complexes, the metal centers are different and will contribute differently to the overall susceptibility. The formalism for treating this question has been worked out nicely by Dunham et al. [45] in their study of the temperature dependence of the shifts in reduced two-iron ferredoxin and can be adapted easily to the semimetHr problem. The temperature dependence of the shifts in semimetHrN$_3$ gives a J value of -20 cm^{-1} following this analysis.*

The substantial decrease in the antiferromagnetic coupling in going from metHrN$_3$ to semimetHrN$_3$ suggests the alteration of the

---

*This value differs from that reported in [27] by virtue of the use of more data and an improved analysis procedure.

bridging unit which mediates the coupling. A simple modification
would be the protonation of the oxo bridge. This suggestion is consistent with observations on the diferric complexes, $[(HBpz_3Fe)_2O-$
$(OAc)_2]$ and its protonated derivative $[(HBpz_3Fe)_2OH(OAc)_2]^+$, which
exhibit J values of -122 and -17 cm^{-1}, respectively [37,38], and on
the diferrous complex $[(Me_3tacnFe)_2OH(OAc)_2]^+$, which exhibits a J
value of -13 cm^{-1} [46]. Though the model compounds are either
[Fe(III), Fe(III)] or [Fe(II), Fe(II)] complexes, it seems unlikely
that [Fe(III), Fe(II)] complexes would deviate significantly from
this pattern. Ongoing studies on other semimetHr complexes [44]
suggest that the hydroxo bridge is characteristic of all semimetHr
species except for the sulfide complex.

The persistence of the carboxylate bridges in semimetHrN$_3$
cannot be confirmed with the NMR data. Nonexchangeable resonances
are observed in the 20 ppm region, which could possibly be assigned
to carboxylate CH_2 protons. However, definitive assignments require
further work on the complex and comparisons with suitable model compounds not yet available.

4.3. Deoxy Forms

Deoxyhemerythrin is known to contain two high-spin Fe(II) centers in
the active site. The x-ray crystallographic information currently
available suggests that few structural differences exist between
oxyHr and deoxyHr [47]. EXAFS analysis of deoxyHr indicates the
absence of an oxo bridge; it is not clear whether the two iron centers remain bridged from this data [48]. The NMR spectrum of deoxyHr
exhibits solvent exchangeable resonances at 45 and 63 ppm (Fig. 5),
assigned to histidine N-H's [28].

The wide range of shifts observed indicates a significant
dipolar component to the isotropic shift and the large shifts observed
indicate the absence of a strong antiferromagnetic interaction.
Indeed, Reem and Solomon recently demonstrated that the high-spin
ferrous centers in deoxyHr are weakly coupled with a J = -13 ± 5 cm^{-1}

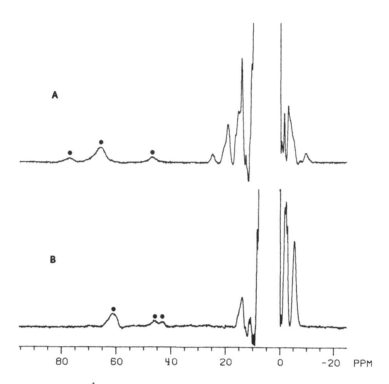

FIG. 5.   300-MHz ^{1}H NMR spectra of deoxyhemerythrin complexes at
45°C in 50 mM phosphate pH 7.5.   (a) azide complex, (b) native form.
A ● denotes a solvent-exchangeable resonance.   (Reprinted with permission from *J. Am. Chem. Soc.*, *107*, 3382 (1985) [28].)

[49].   We are unable to confirm this from a temperature dependence
of the isotropic shifts because the significant dipolar component
to the shift renders such an analysis less than reliable.

It has been demonstrated that the addition of some anions to
deoxyHr affects its reactivity [50].   Indeed azide and cyanate are
proposed to bind to the deoxy binuclear site on the basis of autoxi-
dation experiments [41].   NMR spectroscopy has been useful in providing evidence for this.   The addition of N_3^- to deoxyHr effects
changes shown in Figure 5, with an apparent shift of a 45 ppm feature to 78 ppm.   Changes are also observed in the nonexchangeable
resonances near 20 ppm.

Reem and Solomon showed via MCD and EPR studies that the binding of N_3^- effectively uncouples the two ferrous centers [49]. The larger average isotropic shift observed for deoxyHrN$_3$ N-H resonances is consistent with decreased antiferromagnetic coupling. Further, Evans' susceptibility measurements comparing deoxyHr with deoxyHrN$_3$ demonstrate a substantial increase in χ upon addition of N_3^-. Assuming that deoxyHrN$_3$ is essentially uncoupled, the χ values obtained suggest a $J = -13 \pm 3$ cm^{-1} for the binuclear unit in deoxyHr, in agreement with the MCD results.

Recently, Wieghardt and coworkers [46] reported the synthesis of a binuclear Fe(II)-Fe(II) complex with bridging hydroxo and acetato groups, $[(Me_3tacnFe)_2OH(OAc)_2]^+$. This complex has a structure similar to that of metHr, despite the difference in iron oxidation state, and is proposed as a model for deoxyHr. $J = -13$ cm^{-1} for this complex in agreement with measurements on deoxyHr.

4.4. Sulfide Derivatives

The addition of sulfide to metHr results in the formation of semi-metHrS via a two-step process consisting of reduction to the mixed-valent state and sulfide incorporation into the cluster [51]. Semi-metHrS exhibits a visible absorption band with a maximum near 500 nm and EPR signals at g = 1.89, 1.71, and 1.40. The visible spectrum suggests the presence of S^{2-}-to-Fe(III) charge transfer transitions, while the EPR spectrum is indicative of an antiferromagnetically coupled [Fe(III)-Fe(II)] unit [51,52]. Upon treatment with $[Fe(CN)_6]^{3-}$, semimetHrS is converted to metHrS, which exhibits similar but more intense visible transitions. Both semimetHrS and metHrS show Raman features assigned to Fe-S-Fe vibrations, suggesting that sulfur has replaced the oxygen bridge [51,53]. Neither complex shows a tendency to bind external anions.

The NMR spectrum of metHrS [27] differs little from that of metHr (Fig. 3). The exchangeable NH resonances are resolved into two features and the isotropic shifts are slightly larger. The nonexchangeable feature at 11 ppm assigned to CH_2 protons on carboxylate ligands persists. All this indicates that the replacement of the oxo group with a sulfide has not dramatically altered the structure of the binuclear cluster in the met form. The similarity in the isotropic shifts observed indicate that J is not significantly different. This is consistent with recent studies comparing [Fe(salen)]$_2$O and [Fe(salen)]$_2$S, which show only small differences in the value of J [54].

The NMR spectrum of semimetHrS (Fig. 4) is somewhat more interesting, with the observation of five sharp exchangeable features in the 20 to 55 ppm range. The large number of features observed suggest a significant dipolar contribution to the isotropic shift or the presence of more than one form of the complex in solution. The magnitude of the shifts relative to that observed for semimetHrN$_3$ indicates that the coupling between metal centers is somewhat greater. Indeed, a temperature dependence study of these shifts reveals only small dependences [44]. This lack of a temperature dependence suggests that χ in this temperature range must be near the Néel point. J can thus be estimated to -35 ± 10 cm^{-1}. The larger J value for semimetHrS relative to that of semimetHrN$_3$ probably reflects the differing abilities of a sulfide and an hydroxide to mediate coupling between Fe(III) and Fe(II) centers.

4.5. Overview

The NMR studies discussed above together with other spectroscopic studies strongly indicate that the binuclear cluster in hemerythrin remains intact in the different oxidation states and ligated forms. The single-atom bridge can be modified by protonation as in the cases of semimetHrN$_3$ and deoxyHr or by sulfide replacement as in the cases of metHrS and semimetHrS without leading to the disintegration of the cluster. The picture that thus develops is that of a cluster that can undergo redox reactions with little change in

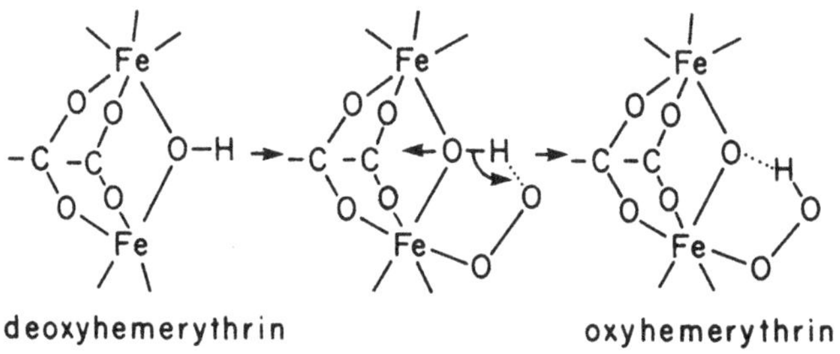

FIG. 6. Oxygenation scheme for hemerythrin.

structure and thus facilitates reversible dioxygen binding. One
such mechanism has been proposed by Stenkamp et al. [47] and shown
in Figure 6.

5. RIBONUCLEOTIDE REDUCTASE

Ribonucleotide reductase catalyzes the deoxygenation of ribonucleo-
tides to deoxyribonucleotides and as such is a key enzyme in DNA
synthesis [55]. The enzyme from *E. coli* consists of two proteins,
B_1 and B_2. The B_2 protein contains a binuclear iron(III) cluster
characterized by intense near-UV absorption, strong antiferromagnetic
coupling ($J = -100$ cm^{-1}) of the iron atoms [56], and the presence of
Fe-O-Fe vibrations in the Raman spectrum [57]. The similarity of
these features to those of methemerythrin has led to suggestions
that these proteins may share the same binuclear cluster structure.

What distinguishes the B_2 protein of ribonucleotide reductase
from methemerythrin is the presence of a tyrosine radical cation
[58]. This radical is essential for enzyme activity and the iron
cluster appears to be involved in the generation and stabilization
of this radical. The radical can be reduced by hydroxylamine and
can be regenerated only by making the apoprotein and treating this
with Fe(II) and O_2 [55].

The NMR spectrum of the native B_2 protein has recently been
reported [29]. There is a solvent exchangeable feature at 24 ppm,

at a position nearly identical to that found in methemerythrin [27].
This suggests that imidazoles are involved in the coordination of
the cluster. There is also a nonexchangeable feature at 19 ppm;
this differs from that observed in methemerythrin at 11 ppm, which
has been assigned to the CH_2 protons of the bridging carboxylates.
Since the antiferromagnetic coupling in the two proteins are comparable and the hemerythrin assignment agrees well with model studies,
it seems possible that the 19 ppm feature in ribonucleotide reductase
may arise from protons on ligands other than bridging carboxylates.
Consequently, other ligands to the iron centers should be considered
in subsequent studies of ribonucleotide reductase.

6. UTEROFERRIN

Uteroferrin belongs to a class of enzymes called purple acid phos-
phatases [59]. Such enzymes have been purified from porcine uterine
fluid, bovine and rat spleen, and hairy cell leukemia cells from
human spleen [60]. They are characterized by a visible absorption
spectrum with an absorbance maximum near 550 nm ($\varepsilon \sim$ 4000 $M^{-1} cm^{-1}$)
and are catalysts for the hydrolysis of phosphate esters. These
enzymes are yet another example of proteins with binuclear iron
active sites.

Uteroferrin can be found in two oxidation states. The oxidized
form is purple, EPR-silent, and enzymatically inactive. Mössbauer
studies show that the iron centers are in the high-spin ferric state
[61]; the lack of an EPR signal and the absence of magnetic hyperfine
Mössbauer features at liquid helium temperatures suggest that the two
irons interact antiferromagnetically. This has been confirmed with
magnetic susceptibility measurements showing a value for J of $\sim$-100
cm^{-1} [62,63].

The reduced form, derived from the addition of an electron, is
pink (λ_{max} 510 nm, $\varepsilon \sim$4000 $M^{-1} cm^{-1}$) and enzymatically active. It
exhibits an EPR spectrum with g values at 1.9, 1.74, and 1.5 [64],
features reminiscent of the semimethemerythrins. The EPR signals

persist up to 40 K but disappear above this temperature, indicating
a short T_{1e}. The low value for g_{ave} (1.74) arises from the S = 1/2
level of an antiferromagnetically coupled high-spin Fe(III)-Fe(II)
pair and Mössbauer studies confirm this description [61].

The properties of the EPR spectrum for the reduced form suggest
that relatively sharp NMR features can be expected. Figure 7 shows
a study on porcine uteroferrin [30]. Well-resolved features are
observed for the reduced protein and their intensities correlate
well with the intensity of the EPR signal. These signals decrease
in intensity as peroxide is added to convert the protein to its oxidized form. No NMR signals outside the diamagnetic region are
observed for the oxidized protein, apparently due to an unfavorable
cluster T_{1e}.

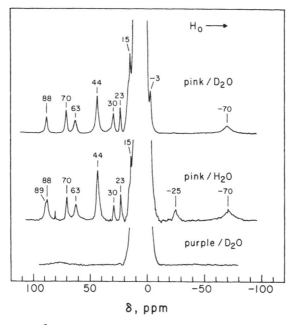

FIG. 7. 300-MHz ^1H NMR spectra of pink and purple uteroferrin in
sodium acetate buffer, pH 4.9, at 30°C. (Reprinted with permission
from *J. Biol. Chem., 258,* 14212 (1983) [30].)

A temperature dependence study of the isotropic shifts of the reduced enzyme allows an estimation of the antiferromagnetic coupling between the iron centers ($J \sim -10$ cm^{-1}). Its similarity to that found for semimethemerythrin azide [27] suggests that hydroxide also bridges the iron centers in reduced uteroferrin.

Some of the NMR features of reduced uteroferrin have been assigned. The visible spectrum has been shown to arise from tyrosinate-to-iron(III) charge transfer transitions by resonance Raman spectroscopy [65,66]. Irradiation into the visible chromophore results in the enhancement of vibrations typical of other iron-tyrosinate proteins [67] such as transferrin and the catechol dioxygenases. Thus, some of the NMR features must arise from tyrosine bound to the iron(III) center.

The characteristic shifts of phenolate protons in iron(III) phenolate complexes have been investigated by Que et al. [31,68]. The phenolate-to-iron(III) charge transfer transition provides the principal mechanism for delocalizing unpaired spin density onto the ligand [68]. Protons in ortho and para positions are shifted upfield, while meta protons are shifted downfield, and to similar extents. The replacement of a ring proton by a methyl group results in the change in the sign of the isotropic shift. These features are diagnostic of a π-delocalization mechanism [2].

In the spectrum of reduced uteroferrin [30], only one feature that is nonexchangeable with solvent is observed in the upfield region (-70 ppm). This is assigned to 3,5-H's of tyrosine (ortho to phenolate O) bound to the ferric center in the mixed-valent cluster. Its position is expected for such protons and its broad linewidth is consistent with its proximity to the metal center. If one considers the potential amino acid ligands, no other Fe(III) ligand would be predicted to have such a large upfield shifted resonance. The corresponding 2,6-H's of tyrosine should be found near +70 ppm and there are two candidates in the spectrum for this assignment (+63 and +70 ppm). Indeed, both resonances can be assigned to those protons because they are related to each other by a dynamic process. Variable-temperature studies show that at 50°C the resonances are in the early stages of coalescence. The dynamic process can be further demonstrated by saturation transfer studies, i.e., saturation of the 63 ppm resonance affects the intensity of the 70 ppm feature. The β-CH$_2$ protons of the coordinated tyrosine are assigned to the resonance at 44 ppm, also on the basis of model studies (Table 3).

The question then arises as to whether tyrosine is also coordinated to the ferrous center. Model studies (Table 3) show that phenolates bound to Fe(II) also exhibit shifts of alternating sign on going from ortho to meta to para. However, these are significantly smaller in magnitude. Examination of the spectrum in the upfield region reveals no nonexchangeable resonance at ~-25 ppm. This leads to the conclusion that tyrosine is not coordinated to the Fe(II) center, i.e., the binuclear cluster is unsymmetric. This conclusion is consistent with two other observations. First, the extinction coefficient of the visible absorption does not change upon oxidation-reduction, suggesting the presence of a chromophoric redox-inactive site and a nonchromophoric redox-active site [59]. Second, only mild reductants are required to convert oxidized uteroferrin to its reduced form, so the redox active site must have a potential near 0 V vs. NHE. The presence of phenolate in the Fe coordination sphere is known to stabilize the ferric state and would lower the potential of the chromophoric site, making it more difficult to reduce.

The other resonances that are readily assignable are the solvent exchangeable features at 89 and 44 ppm. These are consistent with histidine N-H's on imidazoles coordinated to high-spin Fe(III) and Fe(II) centers, respectively. Pulsed EPR studies also suggest the presence of two different nitrogen hyperfine interactions [69], in agreement with the NMR assignment.

The active site structure that develops from these studies features a hydroxo-bridged binuclear iron cluster with tyrosine and histidine ligation at one site and histidine ligation at the other.

Because phenolate-to-iron(III) charge transfer transitions have
extinction coefficients of 1,000-2,000 M^{-1} cm^{-1} per phenolate, it
is likely that two tyrosines are coordinated to the chromophoric
site. Because peak integrations indicate that the number of histidines on each iron corresponds to the number of tyrosines, each iron
would each be coordinated by two histidines. The ligands needed to
complete the iron coordination spheres are currently unknown.

Remaining resonances at 88, 30, 23, and -25 ppm have yet to
be assigned. A comparison of the native reduced uteroferrin spectrum
with those of the molybdate and tungstate complexes (Table 2) shows
that the solvent exchangeable resonance at 44 ppm and the feature at
88 ppm in the native enzyme show the most significant changes in
shift upon anion binding. Since the 44 ppm resonance has been
assigned to a histidine N-H bound to the Fe(II) site, the variability of the shift is not too surprising owing to the larger dipolar
contribution to the isotropic shift expected of a high-spin ferrous
center. The binding of anions may alter the anisotropy of the site,
resulting in the changes observed. Similar arguments would apply to
the resonance at 88 ppm; thus it probably arises from a ligand bound
to the Fe(II) site but the exact assignment will depend on future
experiments.

7. FERREDOXINS

The two-iron ferredoxins represent the first thoroughly studied class
of binuclear iron proteins [70]. Ferredoxins serve as electron transfer agents in many important biological processes, hence the interest
in their structure and function. The nature of the binuclear cluster
in two-iron ferredoxins is well established from crystallographic
[71] and spectroscopic [70] studies on the proteins and from studies
on synthetic analogs [72]. The cluster consists of two tetrahedrally
coordinated iron centers bridged by two sulfides and terminally
ligated to two cysteines. Two oxidation states are accessible:
an oxidized Fe(III)-Fe(III) state and a reduced Fe(III)-Fe(II) state.

Both states are characterized by strong antiferromagnetic coupling
($J_{ox} \sim -182$ cm^{-1} and $J_{red} \sim -80$ to -110 cm^{-1}) [70] and the reduced
state exhibits the now familiar $g = 1.94$ EPR spectrum.

NMR studies on the oxidized form reveal isotropically shifted
features at 32 ppm, assigned to the β-CH$_2$ protons of the coordinated
cysteines [73]. The reduced form exhibits a more complicated spectrum with many resonances between 20 and 120 ppm [73-77]. A temperature dependence study of these resonances shows that some signals
have Curie dependences while others have anti-Curie dependences.
These observations have been modeled well by Dunham et al. [45] in
a formalism that takes into account the relative contributions of
the individual metal centers to the cluster susceptibility in an
antiferromagnetically coupled situation. Put simply, the features
with an anti-Curie temperature dependence can only be ascribed to
protons on the ferrous ligands given the J value found for reduced
ferredoxins. Based on this formalism, Bertini et al. [76] recently
found the long unobserved β-CH$_2$ protons of the ferric cysteinates
at $\sim$120 ppm in spinach ferredoxin. These protons have been particularly difficult to detect because of their large shift and short T_1's.

Other iron-sulfur proteins have also been characterized by NMR
studies [77-81]. At this point, clusters with two, three, and four
iron centers all exhibit distinct NMR properties. However, due to
the complications of the spin-coupling scheme, the NMR spectra at
this point serve only as diagnostic tools for identifying clusters.
As spin coupling in larger clusters becomes better understood, perhaps NMR studies would contribute more basic insights into the structure and bonding of these species.

8. SUMMARY

This chapter has surveyed NMR studies of a number of magnetically
coupled binuclear proteins. In many cases the observation of a
paramagnetically shifted resonance provides a probe into the coordination chemistry of the binuclear cluster, shedding light on the

nature of the ligands in the native state, changes in coordination in other states, and the extent of the antiferromagnetic interaction between the metal centers in the various states. From the examples discussed above, it is clear that the potential applications of this technique for such proteins are just being uncovered. Many techniques, including field-dependent studies of the paramagnetic linewidth, NOE measurements, and specific isotopic labeling experiments, have yet to be applied to these systems and will play a role in future strategies for assigning resonances to specific ligand protons.

ACKNOWLEDGMENTS

This work was supported by the National Science Foundation (DMB-83 14935). L. Que gratefully acknowledges an Alfred P. Sloan Research Fellowship (1982-86) and an NIH Research Career Development Award (1982-87).

ABBREVIATIONS

biim	2,2'-biimidazole
$HBpz_3$	hydrotris(1-pyrazolyl)borate anion
Hr	hemerythrin
im	imidazole
Me_3tacn	1,4,7-trimethyl-1,4,7-triazacyclononane
NHE	normal hydrogen electrode
N-MeTPP	N-methyl-*meso*-tetraphenylporphin
NOE	nuclear Overhauser effect
OAc	acetate
S_2-o-xyl	o-xylene-α,α'-dithiolate
salen	N,N'-ethylenebis(salicylideneamine)
salhis	4-[2'-(salicylideneamino)ethyl]imidazole
salpyr	3-(o-hydroxyphenyl)-5-methylpyrazole
sal_3trien	trisalicylidenetriethylenetetramine

SOD   superoxide dismutase
tacn   1,4,7-triazacyclononane
T(p-isoPrP)P   *meso*-tetra(p-isopropylphenyl)porphin

REFERENCES

1. G. N. LaMar, W. D. Horrocks, Jr., and R. H. Holm (eds.), *NMR of Paramagnetic Molecules: Principles and Applications,* Academic Press, New York, 1973.

2. W. D. Horrocks, Jr., in Ref. 1, p. 128ff.

3. T. J. Swift, in Ref. 1, p. 53ff.

4. G. N. LaMar and F. A. Walker, *J. Am. Chem. Soc., 95,* 6950 (1973).

5. W. Wojciechowski, *Inorg. Chim. Acta, 1,* 319, 324 (1967).

6. (a) C. Benelli, A. Dei, and D. Gatteschi, *Inorg. Chem., 21,* 1284 (1982). (b) A. Dei, D. Gatteschi, and E. Piergentidi, *Inorg. Chem., 18,* 89 (1979).

7. I. Bertini, G. Lanini, C. Luchinat, M. Mancini, and G. Spina, *J. Magn. Reson., 63,* 56 (1985).

8. J. S. Valentine and M. W. Pantoliano, in *Copper Proteins* (T. G. Spiro, ed.), Wiley, New York, 1981, Vol. 3, Chap. 8.

9. J. A. Tainer, E. D. Getzoff, K. M. Beem, J. S. Richardson, and D. C. Richardson, *J. Mol. Biol., 160,* 181 (1982).

10. G. Rotilio, L. Morpurgo, C. Gioragnoli, L. Calabrese, and B. Mondovi, *Biochemistry, 11,* 2187 (1972).

11. L. Calabrese, D. Cocco, L. Morpurgo, B. Mondovi, and G. Rotilio, *Eur. J. Biochem., 64,* 465 (1976). J. A. Fee, *J. Biol. Chem., 248,* 4229 (1973).

12. G. Rotilio, L. Calabrese, B. Mondovi, and W. E. Blumberg, *J. Biol. Chem., 249,* 3157 (1974).

13. I. Bertini, G. Lanini, C. Luchinat, L. Messori, R. Monnanni, and A. Scozzafava, *J. Am. Chem. Soc., 107,* 4391 (1985).

14. I. Bertini, C. Luchinat, and R. Monnanni, *J. Am. Chem. Soc., 107,* 2178 (1985).

15. G. D. Lawrence and D. T. Sawyer, *Biochemistry, 18,* 3045 (1979).

16. T. H. Moss and J. A. Fee, *Biochem. Biophys. Res. Commun., 66,* 799 (1975).

17. D. B. Bailey, P. D. Ellis, and J. A. Fee, *Biochemistry, 19,* 591 (1980).

18. N. J. Blackburn, S. S. Hasnain, N. Binsted, G. P. Diakum, C. D. Garner, and P. F. Knowles, *Biochem. J.*, *219*, 985 (1984).

19. J. Sanders-Loehr and T. M. Loehr, *Adv. Inorg. Biochem.*, *1*, 235 (1979).

20. I. M. Klotz and D. M. Kurtz, Jr., *Accts. Chem. Res.*, *17*, 16 (1984).

21. R. E. Stenkamp, L. C. Sieker, and L. H. Jensen, *J. Am. Chem. Soc.*, *106*, 618 (1984).

22. M. Y. Okamura, I. M. Klotz, C. E. Johnson, M. R. C. Winter, and R. J. P. Williams, *Biochemistry*, *8*, 1951 (1969).

23. J. L. York and A. J. Bearden, *Biochemistry*, *9*, 4549 (1970).

24. K. Garbett, C. E. Johnson, I. M. Klotz, M. Y. Okamura, and R. J. P. Williams, *Arch. Biochem. Biophys.*, *142*, 574 (1971).

25. J. B. R. Dunn, D. F. Shriver, and I. M. Klotz, *Biochemistry*, *14*, 2689 (1975).

26. A. K. Shiemke, T. M. Loehr, and J. Sanders-Loehr, *J. Am. Chem. Soc.*, *106*, 4951 (1984).

27. M. J. Maroney, R. B. Lauffer, L. Que, Jr., and D. M. Kurtz, Jr., *J. Am. Chem. Soc.*, *106*, 6445 (1984).

28. J. M. Nocek, D. M. Kurtz, Jr., J. T. Sage, P. G. Debrunner, M. J. Maroney, and L. Que, Jr., *J. Am. Chem. Soc.*, *107*, 3382 (1985).

29. M. Sahlin, A. Ehrenberg, A. Gräslund, and B. M. Sjöberg, *Rev. Port. Quim.*, *27*, 157 (1985). M. Sahlin, A. Ehrenberg, A. Gräslund, and B. M. Sjöberg, *J. Biol. Chem.*, *261*, 2778 (1986).

30. R. B. Lauffer, B. C. Antanaitis, P. Aisen, and L. Que, Jr., *J. Biol. Chem.*, *258*, 14212 (1983).

31. R. H. Heistand, II, R. B. Lauffer, E. Fikrig, and L. Que, Jr., *J. Am. Chem. Soc.*, *104*, 2789 (1982).

32. H. C. Silvis, Ph.D. thesis, Michigan State University, p. 103, 1981.

33. R. P. Pillai, R. E. Lenkinski, T. T. Sakai, J. M. Geckle, N. R. Krishna, and J. D. Glickson, *Biochem. Biophys. Res. Commun.*, *96*, 341 (1980).

34. A. L. Balch, Y.-W. Chan, G. N. LaMar, L. Latos-Grazynski, and M. W. Renner, *Inorg. Chem.*, *24*, 1437 (1985).

35. H. Goff and G. N. LaMar, *J. Am. Chem. Soc.*, *99*, 6599 (1977).

36. G. N. LaMar, G. R. Eaton, R. H. Holm, and F. A. Walker, *J. Am. Chem. Soc.*, *95*, 63 (1973).

37. W. H. Armstrong, A. Spool, G. C. Papaefthymiou, R. B. Frankel, and S. J. Lippard, *J. Am. Chem. Soc.*, *106*, 3653 (1984).

38. W. H. Armstrong and S. J. Lippard, *J. Am. Chem Soc.*, *106*, 4632 (1984).

39. K. Wieghardt, K. Pohl, and W. Gebert, *Angew. Chem. Int. Ed. Engl., 22,* 727 (1983).

40. J. W. Dawson, H. B. Gray, H. E. Hoenig, G. R. Rossman, J. M. Schredder, and R. H. Wang, *Biochemistry, 11,* 461 (1972).

41. R. G. Wilkins and P. C. Harrington, *Adv. Inorg. Biochem., 5,* 51 (1983).

42. W. O. Gillum, R. B. Frankel, S. Foner, and R. H. Holm, *Inorg. Chem., 15,* 1095 (1976).

43. B. Chiari, O. Piovesana, T. Tarantelli, and P. F. Zanazzi, *Inorg. Chem., 21,* 2444 (1978).

44. M. J. Maroney and L. Que, Jr., unpublished observations.

45. W. R. Dunham, G. Palmer, R. H. Sands, and A. J. Bearden, *Biochim. Biophys. Acta, 253,* 373 (1971).

46. P. Chaudhuri, K. Wieghardt, B. Nuber, and J. Weiss, *Angew. Chem. Int. Ed. Engl., 24,* 778 (1985).

47. R. E. Stenkamp, L. C. Sieker, L. H. Jensen, J. D. McCallum, and J. Sanders-Loehr, *Proc. Natl. Acad. Sci. USA, 82,* 713 (1985).

48. W. T. Elam, E. A. Stern, J. D. McCallum, and J. Sanders-Loehr, *J. Am. Chem. Soc., 105,* 1919 (1983). W. T. Elam, E. A. Stern, J. D. McCallum, and J. Sanders-Loehr, *J. Am. Chem. Soc., 104,* 6369 (1982).

49. R. C. Reem and E. I. Solomon, *J. Am. Chem. Soc., 106,* 8323 (1984).

50. Z. Bradić, R. Conrad, and R. G. Wilkins, *J. Biol. Chem., 252,* 6069 (1975).

51. G. S. Lukat, D. M. Kurtz, Jr., A. K. Shiemke, T. M. Loehr, and J. Sanders-Loehr, *Biochemistry, 23,* 6416 (1984).

52. D. M. Kurtz, Jr., J. T. Sage, M. Hendrich, P. Debrunner, and G. S. Lukat, *J. Biol. Chem., 258,* 2115 (1983).

53. S. M. Freier, L. L. Duff, R. P. VanDuyne, and I. M. Klotz, *Biochemistry, 18,* 5372 (1979).

54. J. R. Dorfman, J. J. Girerd, E. D. Simhon, T. D. P. Stack, and R. H. Holm, *Inorg. Chem., 23,* 4407 (1984).

55. M. Lammers and H. Follman, *Struct. Bonding (Berlin), 54,* 27 (1983). B. M. Sjöberg and A. Graslund, *Adv. Inorg. Biochem., 5,* 87 (1983).

56. L. Petersson, A. Graslund, A. Ehrenberg, B. M. Sjöberg, and P. Reichard, *J. Biol. Chem., 255,* 6706 (1980).

57. B. M. Sjöberg, T. M. Loehr, and J. Sanders-Loehr, *Biochemistry, 21,* 96 (1982).

58. B. M. Sjöberg, P. Reichard, A. Gräslund, and A. Ehrenberg, *J. Biol. Chem., 253,* 6863 (1978).

59. B. C. Antanaitis and P. Aisen, *Adv. Inorg. Biochem.*, *5*, 111
     (1983).

60. C. M. Ketcham, G. A. Baumbach, F. W. Bazer, and R. M. Roberts,
     J. Biol. Chem., *260*, 5768 (1985).

61. P. G. Debrunner, M. P. Hendrich, J. de Jersey, D. T. Keough,
     J. T. Sage, and B. Zerner, *Biochim. Biophys. Acta*, *745*, 103
     (1983).

62. B. C. Antanaitis, P. Aisen, and H. R. Lilienthal, *J. Biol.
     Chem.*, *258*, 3166 (1983).

63. E. Sinn, C. J. O'Connor, J. de Jersey, and B. Zerner, *Inorg.
     Chim. Acta*, *78*, L13 (1983).

64. B. C. Antanaitis, P. Aisen, H. R. Lilienthal, R. M. Roberts,
     and F. W. Bazer, *J. Biol. Chem.*, *255*, 11204 (1980).

65. B. P. Gaber, J. P. Sheridan, F. W. Bazer, and R. M. Roberts,
     J. Biol. Chem., *254*, 8340 (1979).

66. B. C. Antanaitis, T. Strekas, and P. Aisen, *J. Biol. Chem.*,
     257, 3766 (1982).

67. L. Que, Jr., *Coord. Chem. Rev.*, *50*, 73 (1983).

68. J. W. Pyrz, A. L. Roe, L. J. Stern, and L. Que, Jr., *J. Am.
     Chem. Soc.*, *107*, 614 (1985).

69. B. C. Antanaitis, J. Peisach, W. B. Mims, and P. Aisen, *J.
     Biol. Chem.*, *260*, 4572 (1985).

70. G. Palmer, in *Iron-Sulfur Proteins*, Vol. 2 (W. E. Lovenberg,
     ed.), Academic Press, New York, 1973, p. 285ff.

71. T. Tsukihari, K. Fukuyama, M. Nakamura, Y. Katsube, N. Tanaka,
     M. Kakudo, K. Wada, T. Hase, and H. Matsubara, *J. Biochem.
     (Tokyo)*, *90*, 1763 (1982).

72. J. M. Berg and R. H. Holm, in *Metal Ions in Biology* (T. G.
     Spiro, ed.), Wiley, New York, 1982, Vol. 4, Chap. 1.

73. I. Salmeen and G. Palmer, *Arch. Biochem. Biophys.*, *150*, 767
     (1972).

74. J. D. Glickson, W. D. Phillips, C. C. McDonald, and M. Poe,
     Biochem. Biophys. Res. Commun., *42*, 271 (1971).

75. T. M. Chan and J. L. Markley, *Biochemistry*, *22*, 6008 (1983).

76. I. Bertini, G. Lanini, and C. Luchinat, *Inorg. Chem.*, *23*, 2729
     (1984).

77. K. Nagayama, Y. Ozaki, Y. Kyogoku, T. Hase, and H. Matsubara,
     J. Biochem. (Tokyo), *94*, 893 (1983).

78. D. G. Nettesheim, T. E. Meyer, B. A. Feinberg, and J. D. Otvos,
     J. Biol. Chem., *258*, 8235 (1983).

79. K. Nagayama and D. Ohmori, *FEBS Lett.*, *173*, 15 (1984).

80. W. V. Sweeney, *J. Biol. Chem.*, *256*, 12222 (1981).

81. K. Nagayama, D. Ohmori, T. Imai, and T. Oshima, *FEBS Lett.*, *158*, 208 (1983).

4

Proton NMR Studies of Biological Problems Involving Paramagnetic Heme Proteins

James D. Satterlee
University of New Mexico
Department of Chemistry
Albuquerque, New Mexico 87131

1. HEME PROTEINS

1.1. Introduction

Heme proteins are widely distributed in both animal and plant kingdoms and perform several functions which are intimately related to
cellular bioenergetics. As a group, their name derives from the
presence of one or more iron porphyrins, or hemes, in the protein.
Typically, the heme is located at the active site and, as such, is
the focus for the protein's function. There are many different
types of heme structures and nature seems to have tailored different
uses for the different types of hemes (Table 1). The chemistry of
porphyrins alone is rich and varied [1,2]. In combination with
different polypeptide chains even greater diversity of function is
achieved.

It is clear from Table 1 and Figures 1-3 that nature utilizes
the flexibility of different heme structures with various iron and
porphyrin oxidation states. Thus, the iron ion can occur in 2+, 3+,
and 4+ states and under certain circumstances the porphyrin ring
itself may be oxidized or reduced. These features will be discussed
in more detail in later sections and it is sufficient to emphasize
only that the variety of heme protein chemistry is in many cases
attributable to the heme.

It is the purpose of this chapter to provide the reader with
a background to the extant literature concerning the study of paramagnetic forms of heme proteins by proton nuclear magnetic resonance

TABLE 1

Survey of Representative Heme Proteins Indicating the Heme Iron Oxidation
States Observed for Native Proteins, Heme Structure Type, and a
Short Statement of Each Protein's Function and Occurrence

Protein	Native oxid. state	Heme	Function and occurrence[a]	Ref.
Hemoglobin	Fe^{2+}	Protoheme	Oxygen transport in circulatory system	72
Myoglobin	Fe^{2+}	Protoheme	Oxygen storage in muscle	72
Leghemoglobin	Fe^{2+}	Protoheme	Oxygen binding, storage in root nodules	
Peroxidases:				
Cytochrome c	Fe^{3+}	Protoheme	Catalyzes oxidation of ferrous cyt. c	109
Horseradish	Fe^{3+}	Protoheme	Catalyzes oxidation reactions	
Chloro	Fe^{3+}	Protoheme	Catalysis, including chlorination of substrates	
Cytochrome c	Fe^{2+},Fe^{3+}	Heme c	Electron transport in mitochondria	26
Cytochrome c-551	Fe^{2+},Fe^{3+}	Heme c	Electron donor to terminal oxidase in pseudomonads	
Cytochrome c-2	Fe^{2+},Fe^{3+}	Heme c	Electron transfer to oxidase or photosynthetic reaction site	
Cytochrome d	Fe^{2+}	Heme d	Component of terminal oxidase in *E. coli*	133
Cytochrome oxidase	Fe^{2+}	Heme a	Terminal oxidase in electron transport chain, O_2 reduction	
Cytochrome P-450	Fe^{2+},Fe^{3+}	Protoheme	Substrate oxidation; oxygen reduction (for P-450$_{cam}$)	24
Cytochrome b$_5$	Fe^{2+},Fe^{3+}	Protoheme	Electron transfer; methemoglobin reductase system	25,134,135

[a]The function and occurrence of many types of heme proteins is often much broader than indicated in these
summary statements.

(NMR) spectroscopy. This will not be exhaustive, but it is hoped
that it will be sufficiently complete to allow the reader to appre-
ciate the wealth of information available from such NMR studies.
Reviews have appeared elsewhere which deal with aspects of this
subject [4-7] and recently a fairly exhaustive summary of the literature between 1978 and 1984 has appeared [8]. It is also the goal of
this chapter to elaborate on several of the more important techniques
currently being employed as well as to preview emerging areas of
paramagnetic heme protein research.

1.2. Prosthetic Group

Several different heme structures have been identified in heme proteins (Figs. 1 and 2). Nomenclature for hemes has been discussed

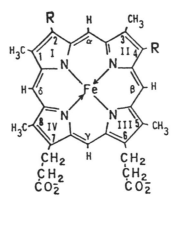

2,4-R	NAME
VINYL	PROTOHEME IX
ETHYL	MESOHEME IX
PROTON	DEUTEROHEME IX
2-PROTON 4-VINYL	PEMPTOHEME
2-VINYL 4-PROTON	ISOPEMPTOHEME

FIG. 1. Structure of the heme group (heme b) indicating by 2,4
substituent the appropriate common name.

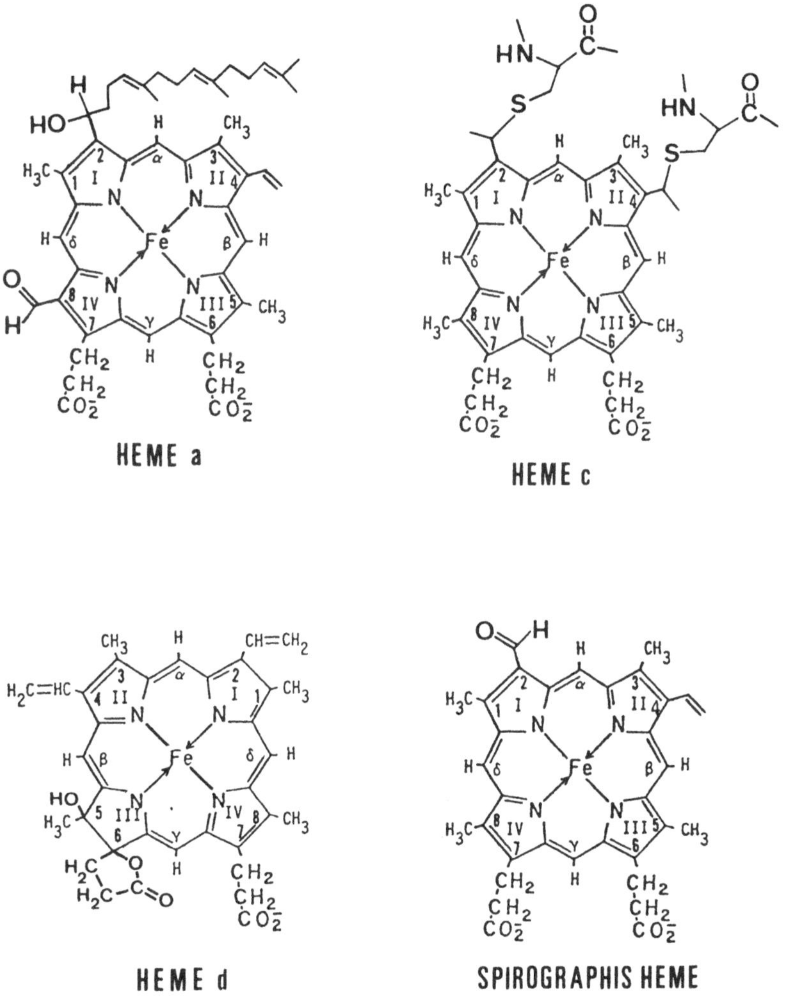

FIG. 2. Structures of several heme derivatives. Heme a is a prosthetic group in the cytochrome oxidases. Heme c occurs in c-type cytochromes. Heme d [133] is found in cytochrome d, which itself is a component of the terminal oxidase complex in *E. coli*. The spirographis heme is frequently called chlorocruoroporphyrin and is the prosthetic group of certain marine annelids, including *S. starte indica*.

previously [2,3]. Both common names and IUPAC rules are typically encountered in the current literature and the reader is referred to [3] for a discussion. Common names will be primarily employed in this chapter.

Heme proteins can be broadly divided into two categories: those that contain a covalent attachment between the porphyrin ring and the polypeptide chain and those that do not. The latter are typified by the b-type heme proteins whose prosthetic group is the iron complex of protoporphyrin IX [also called protoheme(IX) or {protoporphyrinato (IX)}-iron(II, III)]. Although protoheme(IX) is the most frequently encountered b-type heme group, it is worth mentioning that heme a, spirographis porphyrin, siroheme, and heme d (as well as other chlorins), which appear in the green heme proteins, are other heme derivatives that occur in the broader class of non-covalent heme proteins [9-13]. Structures of some of these porphyrins are shown in Figures 1 and 2.

Heme c, the prosthetic group of cytochromes c, is the example of a prosthetic group that appears in nature covalently linked to the protein's polypeptide chain. This structure is also shown in Figure 2. The covalent attachment occurs through the porphyrin 2,4 position via thioether groups donated from polypeptide cysteine side chains.

1.3. Spin, Oxidation, and Coordination States

The most commonly encountered iron oxidation states in natural heme proteins are 2+ (ferrous) and 3+ (ferric). The ferrous state is the normal, functional iron oxidation state for the ligand-binding proteins hemoglobin (Hb) and myoglobin (Mb). For these heme proteins oxidation to the 3+ state results in loss of O_2 and CO ligand-binding ability and such forms are referred to as methemoglobin and metmyoglobin. In contrast, the heme peroxidases occur naturally in the ferric (met) states. In comparison with the heme globins these enzymes are generally unstable with respect to reduction to the ferrous state. Both types of heme proteins just discussed, as well as common cytochromes c, are readily transformed between the ferric and ferrous states employing dithionite ion as a reducing agent and ferricyanide ion as an oxidizing agent. In addition, ascorbate ion is a common reducing agent for the cytochromes c.

Heme proteins that are normally capable of ligand binding include hemoglobins, myoglobins, peroxidases, oxidases, cytochromes P-450, and catalases. For hemoglobins and myoglobins the ferrous state readily binds small neutral ligands such as molecular oxygen and carbon monoxide. In the case of oxygen binding this is accomplished without rapid oxidation to the ferric state. Other heme proteins are also capable of binding carbon monoxide when they are in the ferrous state (cytochrome c peroxidase [14], horseradish peroxidase [15], cytochromes P-450 [16,17], chloroperoxidase [18], myeloperoxidase [19], and lactoperoxidase [20]). However, oxygen binding in the ferrous state of these same proteins is more complicated, resulting in oxoperoxidase (compound III) formation for the peroxidases, which in the cases of cytochrome c peroxidase and horseradish peroxidase ultimately leads to heme iron oxidation back to the ferric state [21-23]. Metastable oxygenated forms of bacterial cytochrome P-450 occur when camphor is present (half-life ~40 min at 4°C) followed, ultimately, by oxidation to the ferric state [17,24].

The ferric states of the heme proteins identified above can also bind ligands. The most commonly encountered ligands are cyanide ion (CN^-), azide ion (N_3^-), and fluoride ion (F^-).

Ligand binding by heme proteins may cause changes in the total electron spin quantum number (Fig. 3). This in turn affects the observed NMR spectrum for reasons that shall become obvious in the next section. The principal coordination structures and iron d-orbital electron configurations for each of the common iron oxidation states (2+, 3+, 4+) encountered in heme proteins are represented in Figures 3 and 4. From these figures it is easy to see that ligand binding may cause spin state transitions (Fig. 3) and accompany structural changes (Fig. 4), or not. For example, the reaction of deoxy hemoglobin with molecular oxygen:

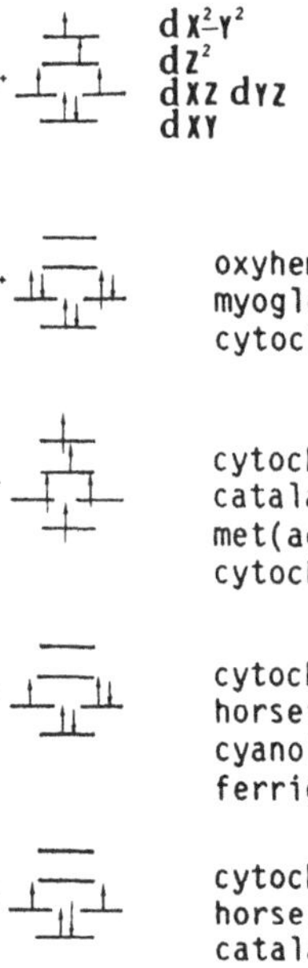

Spin/Ox.   e⁻   Proteins
State   Config.

A   H.S. Fe^{2+} $d\,x^2-y^2$
                          $d\,z^2$
                          $d\,xz\ \ d\,yz$ deoxyhemoglobin; deoxymyoglobin;
                          $d\,xy$ reduced horseradish peroxidase

B   L.S. Fe^{2+} oxyhemoglobin; oxymyoglobin; hemoglobin-CO;
                          myoglobin-CO; reduced cytochrome c;
                          cytochrome P450 hydroxylating intermediate

C   H.S. Fe^{3+} cytochrome c peroxidase; horseradish peroxidase;
                          catalase; cytochrome c peroxidase-fluoride;
                          met(aquo)hemoglobin; met(aquo)myoglobin;
                          cytochrome c'

D   L.S. Fe^{3+} cytochrome c peroxidase-cyanide;
                          horseradish peroxidase-cyanide; cyano(met)myoglobin;
                          cyano(met)hemoglobin; ferric cytochrome c-cyanide;
                          ferric cytochrome b5; cytochrome c'-cyanide

E Fe^{4+} cytochrome c peroxidase compounds I,II;
                          horseradish peroxidase compounds I,II;
                          catalase compound I

FIG. 3. The most common spin and oxidation states of heme proteins are given in this figure along with the electron configuration of each state assuming C_4 symmetry about the heme iron ion. Representative proteins are listed for each case: (A) high-spin (H.S.) ferrous, S = 2; (B) low-spin (L.S.) ferrous, a nonparamagnetic S = 0 form; (C) high-spin ferric, S = 5/2; (D) low-spin ferric, S = 1/2; and (E) the S = 1 ferryl state.

$$Hb + 4O_2 \rightleftharpoons Hb(O_2)_4 \tag{1}$$

results in conversion of each of the hemoglobin subunit hemes from a paramagnetic [Fig. 3(A)], S = 2 state that is structurally five-coordinate (six-coordinate if water occupies an axial coordination position), with the iron ion projecting out of the porphyrin plane toward the proximal histidine [Fig. 4(A)], to a diamagnetic [Fig. 3(B)], S = 0 form, where the iron ion is structurally six-coordinate and lies in the plane of the porphrin ring [Fig. 4(D)].

A similar set of structural and spin state changes occur for cyanide ion ligation to met(aquo)myoglobin. The reaction is

$$Mb^+ + CN^- \rightleftharpoons MbCN \tag{2}$$

and starts with the high-spin (S = 5/2), five-coordinate ferric myoglobin [Figs. 3(C), 4(A)] and produces low-spin (S = 1/2), six-coordinate metmyoglobin cyanide [Figs. 3(D), 4(D)].

In contrast, the reaction of cytochrome c peroxidase (CcP) with fluoride ion

$$CcP + F^- \rightleftharpoons CcP-F \tag{3}$$

results in a very minor spin state change and small but specific structural changes unrelated to the heme iron. CcP is a native ferric protein that is essentially high-spin (S = 5/2) in the absence of fluoride ion. Upon fluoride ion binding the heme iron exhibits characteristics indicating transformation to a more fully high-spin state [Figs. 3(C), 4(A)].

Figure 3(E) indicates the S = 1 spin state that is associated with the ferryl heme structure shown in Figure 4(G). This structure is indicated as the intermediate oxidized form of several peroxidases that result from oxidation of the native (ferric) enzymes by hydrogen peroxide.

These figures (1, 2, 4) also indicate the local heme structure for several types of heme proteins and serve to define the heme ring and amino acid side chain designation system. For the amino acid side chains of the protein polypeptide the biochemical system for position assignments will be used in this work, as shown for the axial ligands histidine [Fig. 4(A)], cysteine [Fig. 4(B)], tyrosine [Fig. 4(C)], and methionine [Fig. 4(E)].

For those interested in this field, familiarity with the inter-relationships between heme protein spin, oxidation, and ligation states is a prerequisite for understanding the basic chemistry of these proteins as well as for interpreting observed spectra.

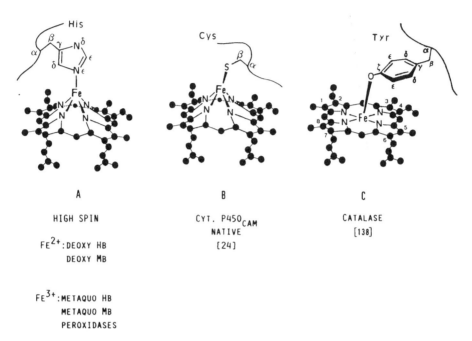

FIG. 4. Representative drawings of the heme coordination structures
in several heme proteins. The porphyrin carbon skeleton is indicated
by black circles. The axial ligands and iron position (in-plane or
out-of-plane) in each case are accurate based on current crystallographic data. Occupation of the sixth coordination site by a water
molecule in the formally five-coordinate species A-C is neglected
due to ambiguity in the structural data in absolutely defining this
aspect of the heme pocket for every protein in this spin and oxidation state. Below each structure are examples of specific protein
forms known to demonstrate each structure and it is to be noted that
similar ligation structures may be adopted by the same protein in
different iron oxidation states. Each axial ligand is identified
in A-G along with the biochemical lettering system that denotes the
carbon-heteroatom side chain.

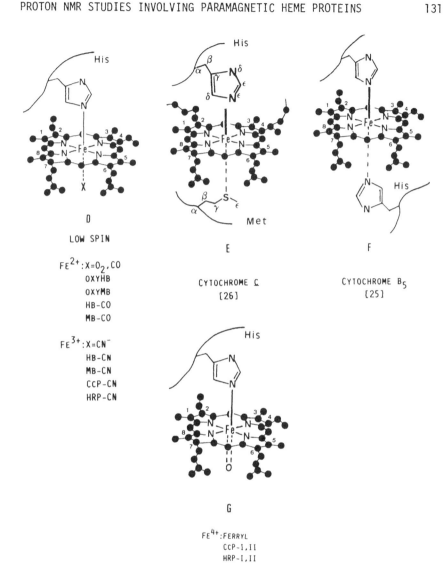

D

LOW SPIN

Fe^{2+}:$X=O_2$,CO
     OXYHB
     OXYMB
     HB-CO
     MB-CO

Fe^{3+}:$X=CN^-$
     HB-CN
     MB-CN
     CCP-CN
     HRP-CN

E

CYTOCHROME C
[26]

F

CYTOCHROME B$_5$
[25]

G

Fe^{4+}:FERRYL
     CCP-I,II
     HRP-I,II

2. THE HYPERFINE SHIFT IN PARAMAGNETIC HEME PROTEINS

2.1. Introduction

Large nuclear shifts may be observed in molecules containing paramagnetic ions. Generally stated, these shifts are a consequence of
the large hyperfine fields that nuclei experience as a result of
the magnetic moments of the free electrons. Not all paramagnetic
complexes demonstrate resolved resonances, however. There are
restrictions on the limits of the unpaired electrons' spin-lattice
relaxation time (T_{1e}) which allow observation of resolved nuclear
resonances. Short T_{1e}'s are required $(T_{1e} \ll \tau_r; T_{1e} < 10^{-11}$ sec)
and this characteristic parameter (T_{1e}) is a consequence of the
orbital ground state electronic configuration of the particular
paramagnetic ion [27]. For this reason, proton NMR resonances of
small free radicals, as an example, are not generally resolvable
$(T_{1e}$ too long) whereas for this same reason metal ions such as Cr^{3+}
and Mn^{2+} are effective relaxation reagents, but complexes of these
metals yield only extremely broad resonances, in most cases too
broad to be observed.

2.2. Relevant Shift Equations

The observed shift in paramagnetic molecules is due to the sum of
diamagnetic and paramagnetic effects. For example, in the case of a
diamagnetic heme, the individual protons of the peripheral substituents exhibit shifts due to differing values of individual diamagnetic
chemical shielding tensors. If the heme becomes paramagnetic, say
with a high-spin ferric iron ion, the large electron magnetic moment
due to the unpaired electrons in the molecule generate large local
magnetic fields at the heme's peripheral protons. Depending on several factors in the following equations, these fields may exert an
effect which can be visualized as supplementing or opposing the
applied field. In one sense this can be thought of as a paramagnetic

shielding or deshielding. If the local fields attributable to the
iron-centered paramagnetism are large enough, an individual proton
may exhibit shifts greater than 100 ppm upfield or downfield from
the reference compounds tetramethylsilane (tms), or 2,2-dimethyl-2-
silapentane-5-sulfonate (dss).

Equations (4) and (5) describe the observed NMR shift in paramagnetic molecules:

$$\delta_{obs} = \delta_{diamagnetic} + \delta_{paramagnetic} \qquad (4)$$

$$\delta_{paramagnetic} = \delta_{contact} + \delta_{dipolar} \qquad (5)$$

where δ refers to the particular shift. The paramagnetic term contains two contributions: the contact shift ($\delta_{contact}$) and the dipolar
shift ($\delta_{dipolar}$). These two terms originate from empirical observations concerning how the magnetic moments of unpaired electrons that
occupy predominantly metal-centered molecular orbitals couple to the
magnetic moments of nuclei, which may be separated from the metal ion
by several intervening chemical bonds. Qualitatively, the dipolar
shift results from the through-space coupling of the magnetic moments
in a manner analogous to the interaction between two separated bar
magnets, whereas the contact shift results from coupling the magnetic
moments via the Fermi contact interaction. The equation describing
the Fermi contact coupling between an electron and a nucleus includes
the expectation value of the electron-containing wave function evaluated at the nucleus, indicating that $\delta_{contact}$ is a consequence of
metal-ligand covalency and the delocalization of unpaired electron
spin density onto the ligand. This can occur be several mechanisms
which have been very well documented for naked porphyrins. The
reviews by LaMar and Walker [28] and Goff [7] are complete and
informative accounts of the studies which have led to the current
picture of valence shell metalloporphyrin bonding that is derivable
from NMR studies.

Equations (6)-(8) give the forms of the contact and dipolar
shift equations that are of most utility to practicing NMR spectro-
scopists. The constants involved in these equations are standard

$$\left(\frac{\Delta H}{H}\right)^{contact} = -\frac{Ag\beta S(S+1)}{3\gamma_N \hbar kT} \tag{6}$$

$$\left(\frac{\Delta H}{H}\right)^{dipolar} = -\frac{\beta^2 S(S+1)}{9kT}(g_\parallel^2 - g_\perp^2)\left(\frac{3\cos^2\theta - 1}{r^3}\right) \tag{7}$$

$$\left(\frac{\Delta H}{H}\right)^{dipolar} = -\frac{1}{2}\left\{[\chi_{zz} - \frac{1}{2}(\chi_{xx} + \chi_{yy})]\left(\frac{3\cos^2\theta - 1}{r^3}\right)\right.$$

$$\left. + \frac{1}{2}(\chi_{xx} - \chi_{yy})\left(\frac{\sin^2\theta\cos 2\Omega}{r^3}\right)\right\} \tag{8}$$

and those not familiar with their origin are referred to other sources for a more complete description [5,8,28,29] as well as to Table 2. In these equations applied to hemes the angle θ is that between the vector linking the iron center with a particular nucleus (for this discussion a proton) and the principal symmetry axis of the molecule (usually the z axis, perpendicular to the heme plane). The angle Ω is that which the projection of this vector on the X-Y plane of an iron-centered cartesian coordinate system makes with the x axis of this system. Note that Eq. (7) is the dipolar shift equation for an axially symmetric system whereas Eq. (8) expresses the equation in terms of the more general rhombic symmetry molecular framework. Equation (7) is most frequently applied due to its simpler form compared with Eq. (8). It works quite well, given the inherent assumptions associated with it, for naked heme and porphyrin complexes where axial molecular symmetry is closely approximated [28]. However, its use in analyzing heme protein spectra is approximate at best and Eq. (8) is the better choice for analyzing paramagnetic dipolar shifts.

In fact, Eqs. (6)-(8) provide only semiquantitative bases for analyzing the observed paramagnetic (isotropic, hyperfine) proton NMR shifts in heme proteins as a result of the many approximations required to bring the more complete equations [28,29] into the forms shown here. It is obvious to those working in this field that the theory of paramagnetic shifts is lacking with respect to precise calculational capability; however, this is a criticism that applies to theories of diamagnetic shifts as well. It is important to understand the limitations and applicability of equations such as (6)-(8)

TABLE 2

Definition of Symbols Used in Eqs. (6)-(12)

Symbol	Definition
χ_{ii}	Principal components of the metal complex magnetic suscepti-bility tensor
θ	Angle between metal-nucleus vector and the principal sym-metry axis (Z) of the heme-centered coordinate system
r	Length of the metal-nucleus vector
Ω	Angle between projection of r on x-y plane and the x axis of the heme-centered coordinate system
S	Electron (total) spin quantum number
β	Bohr magneton
g_{ii}	Epr g values
γ_N	Nuclear magnetogyric ratio
τ_c	$(= \tau_r + T_{1e})$ Dipolar correlation time in the absence of chemical exchange
τ_r	Rotational correlation time (isotropic rotation)
T_{1e}	Electron spin-lattice relaxation time
τ_e	$(= T_{1e})$ Electron hyperfine exchange correlation time in the absence of chemical exchange
T_1	Nuclear spin-lattice relaxation time
T_2	Nuclear spin-spin relaxation time
ω_I	Nuclear Larmor frequency (rad/sec)
ω_S	Electron Larmor frequency (rad/sec)
A	Fermi hyperfine coupling constant
μ_S	Paramagnetic moment

in order to make a critical evaluation of published results. To this end the critical analysis of Jardetzky and Roberts [30] and a review of broader scope than this [8] will provide clarification. That these simpler equations can be successfully applied is emphasized by Keller and Wüthrich's work [31]. This result lends support to the principal assumption employed when qualitative analyses of heme pro-tein hyperfine shift patterns is carried out, namely, that the basic

characteristics which give rise to such shifts in isolated iron
porphyrins are essentially maintained in heme proteins.

2.3. Relaxation Properties

The influence of unpaired electrons upon nuclear spin-lattice (T_1)
and spin-spin (T_2) relaxation times is to shorten both. For example,
normal T_1 values for protons occur in the range of seconds for diamagnetic porphyrins, whereas for ferric low-spin porphyrin proton
resonances T_1 is as low as 10 msec [32]. The reasons for this are
documented [5,27,28,33] and this effect can be described by considering that fluctuations in the electron magnetic moments provide
fluctuating local magnetic fields at nuclei with appropriate spectral
density functions required for efficient nuclear relaxation. Without derivation, the most commonly used equations that describe these
effects are the following:

$$\frac{1}{T_{1,M}} = \frac{2}{15} \frac{\gamma_I^2 g^2 S(S+1)}{r^6} \left(\frac{3\tau_c}{1 + \omega_I^2\tau_c^2} + \frac{7\tau_c}{1 + \omega_S^2\tau_c^2} \right)$$

$$+ \frac{2}{3} S(S+1) \left(\frac{A}{\hbar}\right)^2 \left(\frac{\tau_e}{1 + \omega_S^2\tau_e^2} \right) \tag{9}$$

$$\frac{1}{T_{2,M}} = \frac{1}{15} \frac{\gamma_I^2 g^2 S(S+1)\beta^2}{r^6} \left(4\tau_c + \frac{3\tau_c}{1 + \omega_I^2\tau_c^2} + \frac{13\tau_c}{1 + \omega_S^2\tau_c^2} \right)$$

$$+ \frac{1}{3} S(S+1) \left(\frac{A}{\hbar}\right)^2 \left(\frac{\tau_e}{1 + \omega_S^2\tau_e^2} + \tau_e \right) \tag{10}$$

Complete descriptions of their origin appear elsewhere (see
Table 2 for definitions) [5,8,27-29,30,33-36]. The important aspects
of these equations, insofar as they are applied to heme proteins, are
that both types of interactions that give rise to the hyperfine NMR
shift also affect nuclear relaxation. These are the instantaneous
dipolar coupling between nuclei and unpaired electrons [first term,

Eqs. (9) and (10)] and the hyperfine exchange coupling which occurs
as a consequence of Fermi contact interactions [second term, Eqs.
(9) and (10)]. An additional contribution caused by interaction
between a molecule's rotational motion and the thermally averaged
electron spin magnetic moment of the paramagnetic center has also
been identified. This type of relaxation, given by Eq. (11) for
the same conditions that define the forms of Eqs. (9) and (10),

$$T_2^{-1} \text{(C.S.)} = [\frac{4\gamma_1^2 \mu_S^4 B_0^2}{45 r^6 (kT)^2}] \tau r \tag{11}$$

results in field-dependent linewidths. It is, therefore, an effect
on T_2 [37,38] and it can become the dominant factor at high-field
strengths (8-11T) [39-41]. This effect directly impacts the observability of NMR hyperfine spectra since experience shows that resonances
may broaden severely for higher molecular weight proteins at higher
fields. In cases where Curie spin relaxation is a major contributor
to a nucleus's total relaxation rate, resolution enhancement normally
expected at higher fields may not be realized. The constants that
occur in these equations are defined in Table 2 and other references
[5,8,34-41].

2.4. Oxidation and Spin State Effects on Shifts and Relaxation

Both the oxidation state and the spin state of iron affect the
observed shifts and relaxation times for nuclei of the heme group
and amino acid side chains that lie close to the heme iron center.
Qualitatively, this is obvious from the fact that the equations
presented above, which describe the observed shifts and relaxation
times, all depend on S, the total electron spin quantum number, or
μ_S, the magnetic moment of the paramagnetic center due to the unpaired electrons, and A, the electron-nuclear (Fermi) hyperfine
coupling constant.

Restricting attention to iron porphyrins and heme proteins it
is found that the observed shift pattern of the heme group protons is
characteristic for different spin and oxidation states. For ferrous
hemes the low-spin state is diamagnetic and no proton hyperfine resonances are observed. In the high-spin Fe^{2+} case (S = 2) only small
proton hyperfine shifts are observed for the heme substituents [5-8,
28]. The heme methyl protons are best resolved and they appear in
the range 20-5 ppm (Fig. 5). When proton spectra are collected in
90% protonated water (^1H_2O) rather than in D_2O (^2H_2O), as normally
is the case, exchangeable protons, such as histidine imidazole ring
$N_\delta H$'s, can be observed. Proximal histidine $N_\delta H$ resonances are found

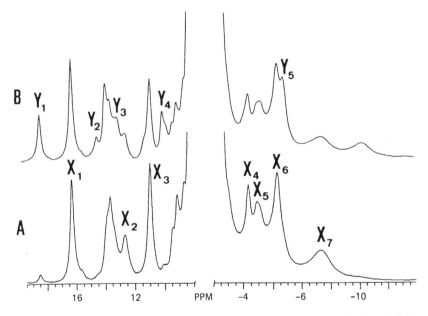

FIG. 5. Proton NMR spectra of sperm whale deoxymyoglobin in 0.2 M
NaCl, 99% ^2H_2O, pH 7.0, 25°C. The scale is in ppm from dss. Trace
(A) native myoglobin with 8% minor component. The major component
peaks are labeled X_1-X_7 and include obvious methyl resonances X_1,X_3.
The minor component peaks are labeled Y_1-Y_5 and are shown to be
present to the extent of 40% in protoheme reconstituted myoglobin,
trace (B). Peak X_1 consists of a methyl as well as a single proton
peak for the major component. (Reproduced from [75] with permission.)

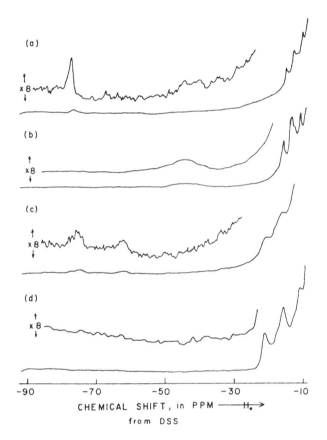

FIG. 6. Proton NMR spectra of deoxy myoglobin and hemoglobin showing the proximal histidine-exchangeable $N_\delta H$ resonances in the region between 55 and 80 ppm downfield. (a) 14 mM sperm whale deoxymyoglobin in 0.2 M NaCl H_2O, pH = 6.8; (b) 9 mM sperm whale deoxymyoglobin in 0.2 M NaCl, 99.8% D_2O, pH meter reading of 7.0; (c) 4 mM deoxyhemoglobin A in 0.2 M NaCl H_2O, pH 6.0; (d) 4 mM deoxyhemoglobin A in 0.2 M NaCl, 99.8% D_2O, pH meter reading of 6.3. Chemical shifts are referenced against internal dss; probe temperature was 25°C. (Reprinted with permission from *Biochemical and Biophysical Research Communications, 77,* 104, copyright 1977, Academic Press.)

between 50 and 80 ppm downfield for ferrous heme proteins [5,6], as shown in Figure 6.

For the high-spin ferric case (S = 5/2), heme methyls are observed between 50 and 90 ppm. Similarly, single protons of the carbon atoms of other pyrrole substituents (vinyl H_α, propionic

acid Hα) also occur downfield, but these demonstrate shifts in the
range 30-60 ppm as shown for cytochrome c peroxidase in Figure 7.
Pyrrole substituent β-carbon protons appear upfield (vinyl H_β,
propionic acid H_β) (Fig. 7). The proximal histidine exchangeable
proton ($N_\delta H$) occurs downfield from 90 ppm for high-spin ferric heme
proteins, as shown for metaquomyoglobin in Figure 8.

Low-spin ferric heme proteins demonstrate a much decreased
proton shift dispersion, extending only over the range between 40 ppm
and -30 ppm at temperatures near 22°C. The heme methyl groups' proton
resonances occur between approximately 0 and 40 ppm with two resonances
normally observable near 35 ppm. For example, in the horse cytochrome

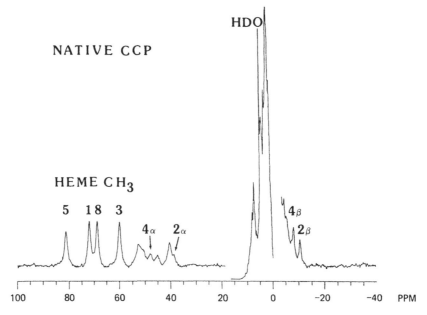

FIG. 7. Proton NMR spectrum at 361 MHz of 2.7 mM native cytochrome c
peroxidase (CcP) taken in 99.8% ^2H_2O (Merck), in 300 mM KNO_3 at 22°C
and pH meter reading 7.0. Shown in this figure are proton resonance
assignments made by reconstituting the native apoprotein with specifically deuterated hemes [55]. The heme methyl groups are shown downfield between 55 and 85 ppm, whereas the α-vinyl proton resonances
are labeled in the downfield region. Two of the four possible β-vinyl
resonances are resolved in the upfield region.

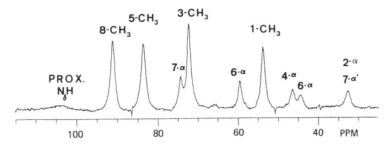

FIG. 8. Proton NMR spectrum at 360 MHz of 3.0 mM metaquomyoglobin (sperm whale) in 90% H_2O at pH 5.0, 23°C with 0.1 M potassium phosphate buffer. Only the downfield hyperfine shift region is shown with the assignments of each resonance that have been reported in [64]. The proximal histidine $N_\delta H$ resonance shown at approximately 103 ppm is absent when this protein is exchanged into an identical D_2O (^2H_2O) solution, whereas the rest of the spectrum is identical to this.

c spectrum shown in Figure 9 heme methyls 8 and 3 are observed downfield whereas heme methyl 5 is found near the broad envelope of overlapping resonances that begins at about 10 ppm. Heme methyl 1 lies upfield, under the envelope. Some of the axial histidine protons also lie downfield as do the α-carbon protons of heme pyrrole substituents (in this case propionic acid substituents, P in Fig. 9). This is also revealed in Figure 10, which gives the spectrum of cyanide-ligated cytochrome c peroxidase. As before, the spectra of Figures 9 and 10 indicate that the β-carbon protons of heme pyrrole substituents lie upfield. In addition, Figure 9 shows that protons of the cytochrome c axial methionine ligand appear far upfield.

The iron 4+ oxidation states of heme proteins, characterized by the oxidized intermediates of the heme peroxidases, demonstrate proton hyperfine spectra that depend on whether the porphyrin ligand is oxidized in the process of generating the ferryl protein [42-47]. When it is not, as in cytochrome c peroxidase compound I [44,45], the product of myoglobin treatment with hydrogen peroxide [46] and the oxidized form of cytochrome c heme peptide (microperoxidase) [47], the only resolved hyperfine resonances that are observed are apparently attributable to the heme meso protons [46]. These resonances

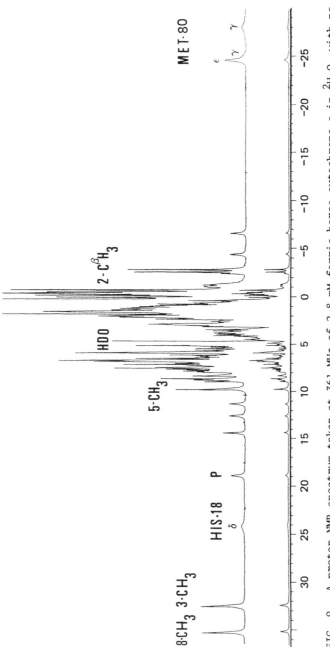

FIG. 9. A proton NMR spectrum taken at 361 MHz of 2.8 mM ferric horse cytochrome c in ^2H_2O, with no added salt or buffer at a pH meter reading of 6.9. The probe temperature was regulated at 24°C. The assignments given include the heme 8,3,5 methyl groups [62]; the axially coordinated histidine (His-18) δ proton, the axially coordinated methionine (Met-91) ε-methyl and γ-methylene resonances are assigned by inference with similar assignments in the yeast cytochromes c [129] and less direct assignments for horse cytochrome c [131]. These should be considered tentative for horse cytochrome c. The 2 position β-methyl group at −2.8 ppm has been assigned by NOE experiments [130]. P indicates assignment of prob-able propionic acid α-methylene protons [131].

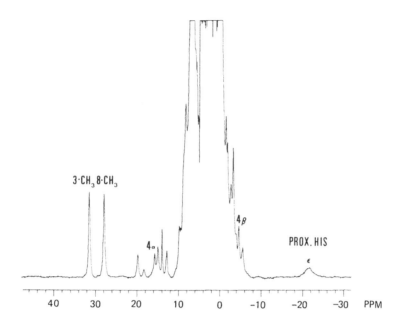

FIG. 10. A proton NMR spectrum at 361 MHz of yeast cytochrome c peroxidase-cyanide in 99.8% ^2H_2O, 300 mM KNO_3, pH meter reading was 7.2 at 22°C. Assignments of heme methyl substituents 3 and 8 and the 4-position vinyl substituents (α-H, β-H) are given [54], as well as the proximal histidine ε proton resonance [132].

lie between 10 and 30 ppm downfield. More detail concerning the heme peroxidase proteins will be presented later.

So far this discussion has focused on the hyperfine resonances that are resolved outside the shift region of approximately -1 to 10 ppm. This region (-1—10 ppm), often referred to as the diamagnetic shift region, is characterized by a broad envelope of overlapping resonances. It has been pointed out that paramagnetic effects are largest in the vicinity of the paramagnetic heme, so that the majority of protons in a protein of reasonable size are little influenced by the heme paramagnetism. Such protons display small differences in their resonance positions compared with their positions in diamagnetic forms of the protein and they occur in this same -1 to 10 ppm region. Analysis of shifts within this region can be profitably employed, as demonstrated elsewhere in this volume, for smaller proteins where resolution of individual resonances is achievable. Furthermore, since the paramagnetic influence can result in shifts that are either downfield or upfield, hyperfine shifted resonances also occur in the -1 to 10 ppm region.

3. RESONANCES OF THE PORPHYRIN RING

3.1. Assignments

The most direct and least ambiguous method for assigning resolved proton hyperfine resonances of the heme group is by reconstituting the apoprotein of interest with protoheme(IX) derivatives for which peripheral substituents have been selectively deuterated. Comparisons of proton NMR spectra of the holoprotein with spectra of the reconstituted protein forms leads to direct assignments of the missing resonances to the specifically deuterated heme groups. This method is available as a result of the tremendous synthetic efforts and structural proofs that have originated in Kevin Smith's laboratory [48,49] and it has been applied to several types of ferric heme proteins [50-56].

As examples of this method our data for proton hyperfine resonance assignments in the predominantly high-spin native cytochrome c peroxidase (CcP) (Figs. 11-13) and the low-spin cyanide ligated form (CcP-CN) (Figs. 14 and 15) are shown [54,55]. In Figure 11 it is seen that compared with the methyl region of the native enzyme, reconstitution with the protoheme(IX) derivative in which the 1,5-position methyl groups are deuterated [Fig. 11(B)] or with the derivative in which the 1,3-position methyl groups are deuterated [Fig. 11(C)] results in loss of intensity of two of the resonances for each case. This leads to direct assignment of the 1,3,5-heme methyl groups' resonances and, by inference, the resonance that remains unchanged throughout this process is assigned to the heme 8-position methyl.

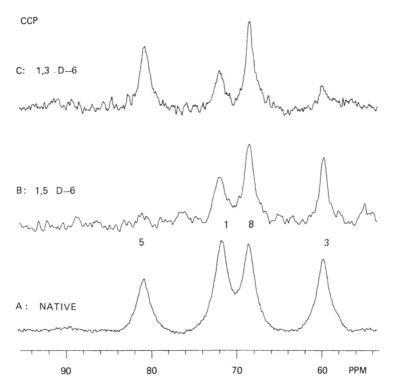

FIG. 11. Proton assignments of the heme methyl groups in cytochrome c peroxidase at 360 MHz, 25°C, and 0.1 M KNO_3. (A) Cytochrome c peroxidase (native), pH 7.5; (B) [1,5-$^2H_6$] hemin-reconstituted cytochrome c peroxidase, pH 7.2; (C) [1,3-$^2H_6$] hemin-reconstituted cytochrome c peroxidase, pH 7.5. (From reference 55, reprinted with permission.)

This process can be repeated for assignments of other heme substituent proton resonances as illustrated in Figures 12-15. In principle, reconstituting b-type heme proteins with iron porphyrins related to protoheme(IX) could also provide a means for assigning the resonances of protons of heme peripheral substituents. For example, deuteroheme, the derivative of protoheme(IX) in which the 2,4-position vinyl groups (Fig. 1) are substituted by protons; could provide assignments of heme vinyl proton resonances. An example

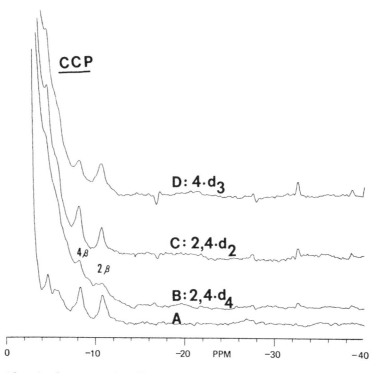

FIG. 12. Assignment of upfield single-proton resonances for cytochrome c peroxidase at 360 MHz, 25°C, 0.1 M KNO_3, pH 7.1-7.4.
(A) native peroxidase; (B) $[2,4-\beta-^2H_4]$ hemin cytochrome c peroxidase;
(C) $[2,4-\alpha-^2H_2]$ hemin cytochrome c peroxidase; (D) $[4-^2H_3]$ hemin
cytochrome c peroxidase. (From reference 55, reprinted with permission.)

from our work with cytochrome c peroxidase indicates that complications can occur [57]. That this may not be a straightforward procedure when the nature of heme peripheral substituents is altered is
shown for deuteroheme-reconstituted CcP-CN in Figure 16. From work
on low-spin ferric complexes of deuteroheme dimethyl ester it is
known that the 2,4-position proton resonances occur upfield between
-10 and -50 ppm. Two such resonances are normally expected. For
deuteroheme-reconstituted CcP-CN one observes five single-proton

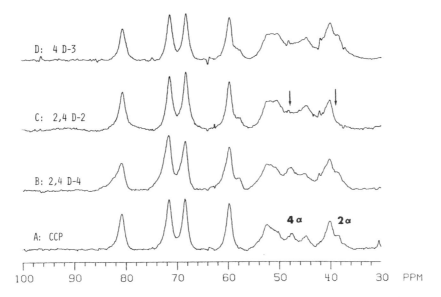

FIG. 13. Single heme proton assignments for cytochrome c peroxidase at 360 MHz, pH 7.1-7.5, 25°C. Downfield proton hyperfine shift region between 32 and 56 ppm, of cytochrome c peroxidase reconstituted with various deuterated hemins: (A) cytochrome c peroxidase; (B) $[2,4-\beta-^2H_4]$ hemin cytochrome c peroxidase; (C) $[2,4-\alpha-^2H_2]$ hemin cytochrome c peroxidase; (D) $[4-^2H_3]$ hemin cytochrome c peroxidase. (From reference 55, reprinted with permission.)

resonances in the upfield region. Moreover, instead of observing two heme methyl resonances downfield between 20 and 35 ppm as CcP normally demonstrates, four heme methyl resonances are detected. Combined with other data, the conclusion to be drawn is that more than one form of CcP-CN exists when deuteroheme is substituted for protoheme. This illustrates an important precaution required by reconstitution studies, even reconstitution studies involving only protoheme(IX) (see Fig. 5 [75]). One must determine whether reconstitutions can be carried out so as to generate a reconstituted protein that is in an identical form to the native protein.

Another method for assigning proton resonances is the nuclear Overhauser technique [4]. This method can be applied in steady-state,

CCP – CN

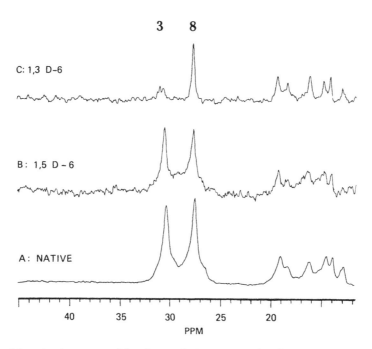

FIG. 14. Assignment of hemin methyl protons by deuteration: (A) native CcP-CN; (B) CcP reconstituted with $[1,5-^{2}H_6]$ hemin; (C) CcP reconstituted with $[1,3-^{2}H_6]$ hemin. The enzyme was in 0.1 M KNO_3, pH 7-7.4, at 24°C. (From reference 54, with permission.)

transient, and truncated driven methods and was pioneered by Wüthrich and his coworkers [58-61]. It is described in greater detail in other reviews [4,8]; however, its results are impressive. For example, the studies on many c-type cytochromes that have come from Wüthrich's laboratory have resulted in identification of two possible configurations for the axial methionine. These orientations, reflecting different chirality about the methionine sulfur, are shown in Figure 17. The distinction between these two possible orientations in solution derives from observed homonuclear proton Overhauser enhancements in

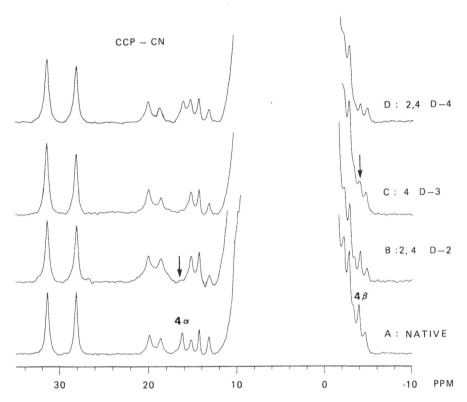

FIG. 15. Assignment of hemin vinyl protons by deuteration in CcP-CN.
Both upfield and downfield regions are shown in this figure. (A)
native CcP-CN; (B) α-vinyl's deuterated; (C) 4-vinyl group per-
deuterated (both α and β protons); (D) β-vinyl protons deuterated.
(From reference 54, reprinted with permission.)

the ferrous proteins [58,62,63]. Thus, for a protein with the axial

methionine orientated as in Figure 17(A), irradiation of the methio-

nine ε-CH_3 resonance is expected to produce differential intensity

enhancements for the α- and δ-meso protons of heme ring compared with

the β- or γ-meso protons. This is precisely what is observed for

horse cytochrome c [63]. In contrast, for cytochrome c-551 from

Pseudomonas aeruginosa [63] irradiation of the axial methionine

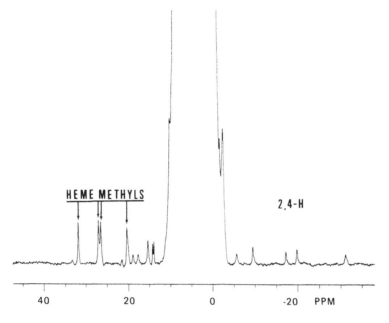

FIG. 16. A proton NMR spectrum at 361 MHz of deuteroheme reconstituted cytochrome c peroxidase (CcP) in 99.8% ^2H_2O with pH meter
reading of 6.94. The protein was in a 0.1 M KNO_3 solution and the
spectrum was accumulated at 22°C. The heme methyl resonances are
shown in the downfield region and the 2,4-position proton resonances
are indicated as the single-proton intensity resonances between -5
and -33 ppm.

methyl resonance results in differential nuclear Overhauser enhancements for the γ-, δ-meso proton resonances compared with the α-, or
β-meso protons. The impact of these kinds of experiments for interpreting the NMR spectrum of paramagnetic forms of these cytochromes
is shown in Table 3 in conjunction with Figure 9. The typical ferric
cytochrome c spectrum (Fig. 9) reveals two heme methyl resonances
shifted the farthest downfield. For horse cytochrome c these are
assigned to the heme 8-CH_3 and 3-CH_3 pair. Table 3 illustrates that
downfield 8,3-methyl resonances correlate with R chirality [Fig.
17(a)] about the axial methionine. In contrast, appearance of heme

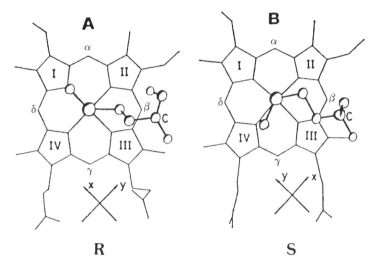

FIG. 17. Perspective computer drawing of the heme group and axial methionine in (A) tuna ferricytochrome c and (B) *P. aeruginosa* ferri-cytochrome c-551. Tuna cytochrome c was selected as a representative example of the group of mammalian-type cytochromes c, which also includes horse cytochrome c, that demonstrate R chirality about the coordinated sulfur atom. Part (B) demonstrates S chirality. The view is in the direction perpendicular to the heme plane. In the R configuration the methionine γ-CH_2 protons are closer to the ϵ-CH_3 than are the β-CH_2 protons. In the S configuration the β-CH_2 group lies closer to the ϵ-CH_3. The directions of the principal axes x and y of the electronic g tensor are also shown. (Reproduced from [63] with permission.)

5-CH_3 and 1-CH_3 resonances downfield correlates with S chirality of the axial methionine [Fig. 17(B)]. In order to explain this feature of the proton spectrum it is necessary to digress momentarily for a brief discussion concerning the cause of this pairwise grouping of heme methyl proton resonances. However, before this is done another comment is required.

Recently, it has been shown that even in high-spin ferric heme proteins direct nuclear Overhauser effects can be observed for the hyperfine shifted resonances [64]. Although these effects are small

TABLE 3

Correlation of Coordinated Axial Methionine Chirality with
the Heme Methyl Group Resonances Exhibiting the Largest
Hyperfine Shift in Selected c-Type Ferric Cytochromes

Protein	Chirality	Assignment of downfield resonances	Ref.
R. rubrum c	R	8,3	60
horse c	R	8,3	63
C. krusei c	R	8,3	129
R. rubrum c2	R	8,3	60
D. vulgaris c553	R	8,3	136
D. sulfuricans c553	R	8,3	136
P. stutzeri c551	S	5,1	60
P. mendocina c551	S	5,1	60
P. aeruginosa c551	S	5,1	63

(2-7%), careful technique has allowed the complete assignment of heme
peripheral substituent proton resonances in metmyoglobin (see Fig. 8)
using one-dimensional NOE connectivities. This is an extremely significant development and defines what undoubtedly will become the
preferred future assignment method.

3.2. Axial Ligand Effects on Hyperfine Shift Patterns

It is known from many studies [5,7,8,53,62,63,65-69] that the proton
hyperfine shift patterns of low-spin ferric heme proteins reflect
pronounced C_2 symmetry in reference to the heme structure. Accordingly, heme methyl group resonances are grouped pairwise (vide supra).
Those pairs are 8,3 and 5,1 so that methyls on heme pyrroles oriented
180° opposite are grouped together. Such shift patterns reflect asymmetry in the electronic wavefunction that contains the unpaired spin
density, which causes the observed hyperfine shifts [5,8,28,54].

This in-plane asymmetry of the hyperfine shift pattern has been
attributed to two effects. As seen above for the various cytochromes
c, axial methionine orientation can regulate the hyperfine proton
resonance pattern. The proximal histidine's imidazole plane can also
affect the shift pattern by virtue of its orientation projected onto
the heme plane [68,70,71]. A prime example of this effect is demonstrated by a comparison of the hyperfine shift patterns of cytochrome
c peroxidase-cyanide (CcP-CN) and metmyoglobin-cyanide (Mb-CN), with
their crystal structures. The hyperfine resonance positions for
these two proteins is shown in Table 4. Notice that the 3,8 methyl
pair exhibits the largest downfield shifts in CcP-CN, whereas in
Mb-CN it is the 5,1 pair which demonstrates the largest downfield
shifts. Figure 18 indicates the proximal imidazole's plane orientation projected onto the heme for both CcP and Mb. For CcP, pyrroles
II (3-CH$_3$) and IV (8-CH$_3$) lie perpendicular to the imidazole plane
projection and it is these methyl groups that exhibit the largest
downfield shifts. For myoglobin, pyrroles I (1-CH$_3$) and III (5-CH$_3$)

TABLE 4

Methyl Hyperfine Resonance Positions
for Ferric Heme Proteins

		Heme methyls			Ref.
Mb-CN					
rel order	5	1	8	3	
obsd shift (25°C)	27.4	18.7	13.0	b	137
pyrrole	III	I	IV	II	
CcP-CN					
rel order	3	8			
obsd shift (22°C)	31.2	27.3	b	b	54
pyrrole	II	IV		(III,I)	
HRP-CN					
rel order	8	3	b	b	
obsd shift (35°C)	29.8	24.9			*)
pyrrole	IV	II		(III,I)	

^{a}In ppm.
^{b}Resonances not resolved.

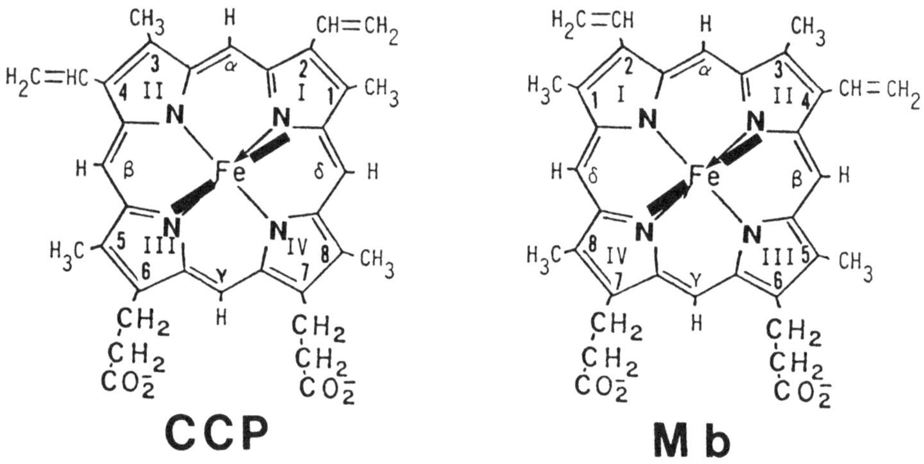

FIG. 18. Comparison of the proximal histidine (imidazole) plane
projection onto the hemin for CcP and Mb. This view is along the
proximal histidine-Fe bond and was formulated from the published
crystal structures for each. Note the interchange of pyrroles I,II
and III,IV in the two structures. This is equivalent to a 180°
rotation about the α-γ (meso) axis. (From reference 54, reprinted
with permission.)

lie perpendicular to the imidazole plane projection and the methyl
groups of these pyrroles lie farthest downfield. The bonding scheme
required to account for this effect involves interaction of the
porphyrin and imidazole π orbitals with iron d π orbitals as previously described [28,68,70,71]. In both the cytochromes c and
several b-type heme proteins the experimental results indicate that
the iron axial ligands are capable of determining the hyperfine shift
pattern. Since it has been concluded that this shift pattern reflects
the distribution of unpaired spin density about the porphyrin macrocycle (contact shift), these heme axial ligands are capable of altering a protein's heme electron delocalization pattern, a fact that may
be important for heme protein redox partners (cf. Sec. 6).

3.3. Heme Contacts with the Protein

Peripheral contacts between amino acid side chains of the protein
polypeptide and the heme may also influence the observed hyperfine
shift pattern [66,68]. In fact, differences in heme peripheral contacts for CcP-CN and horseradish peroxidase-cyanide (HRP-CN) have
been interpreted as a secondary effect that reverses the ordering
of the downfield 8,3-heme methyl resonances for these two proteins.
This is shown in Table 4. Because the pyrrole II and IV substituents
lie downfield for both peroxidases, the proximal histidine orientation
must be essentially as shown in Figure 18(a) for both CcP-CN and
HRP-CN. However, the 3-CH_3 lies farthest downfield for CcP-CN whereas
the 8-CH_3 lies farthest downfield for HRP-CN. The ordering reversal
has been correlated with a specific primary sequence amino acid difference for these two proteins which alters the heme-protein contact
[54].

3.4. Heme Substituents as Regulators of Heme
Pocket Geometry and Ligand Binding

The advent of x-ray crystallography applications to heme protein
research provides wonderful opportunities to those interested in
biological structures and the correlation of structure with function.
However, one must continually be conscious of the fact that the rigid
structures presented in publications [72] are incapable of communicat-
ing the fact that biological molecules are generally dynamic structures. This has been pointed out by Karplus and his coworkers, who
examined the amino acid side chain fluctuations required for ligand
binding to heme proteins [73,74]. Furthermore, crystallography was
not capable of detecting the two heme-polypeptide orientation isomers
that NMR has shown naturally exist for sperm whale myoglobin [75-77],
or the *Chironomous thummi thummi* monomeric hemoglobins [78-81].
These heme orientation isomers exist as a consequence of considering
the heme rotated by 180° about its α-γ meso axis (Fig. 1) and placing

it into the fixed-protein matrix in either of these two orientations. This description does not mean to imply a mechanism for the establishment of heme-centered asymmetry. It is merely a description of the geometric differences of the two forms. The nuclear Overhauser effect (NOE) method has recently been applied in an elegant manner to confirmation of this picture in the case of sperm whale Mb-CN [77]. It is particularly important for this case in view of the fact that the two heme-based isomeric forms of reduced myoglobin exhibit significantly different oxygen affinities [82]. This work implies that ligand-binding dynamics are affected by the peripheral contacts between a heme and its protein matrix.

Figures 19-22 [77] reveal how NOE experiments can establish the heme orientation in ferric low-spin heme proteins. Figure 19(A)

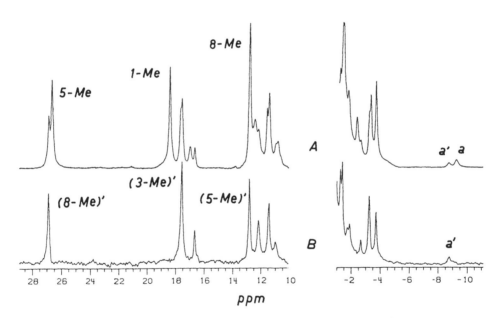

FIG. 19. (A) 500-MHz proton spectrum of freshly reconstituted sperm whale metcyanomyoglobin in ^2H_2O, 0.2 M in NaCl at 30°C and pH* 8.6. A prime superscript indicates a peak arising from the minor form. (B) Spectrum of the minor component calculated from (A) and a spectrum of the same sample recorded after partial equilibration. The later spectrum was scaled so that in the difference the major peaks are removed. (Reproduced from [77] with permission.)

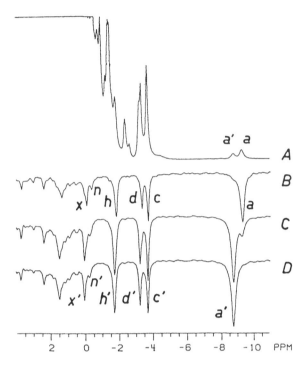

FIG. 20.   (A) 500-MHz proton spectrum of freshly reconstituted sperm whale metcyanomyoglobin in ^2H_2O 0.2 M NaCl at 30°C and pH* 8.6; region upfield from the residual H^2HO signal.   (B) Nuclear Overhauser effect difference spectrum resulting from saturation of peak (a), intensity x5.   (C) Nuclear Overhauser effect difference spectrum resulting from saturation of peak (a'), intensity x10.   (D) Same as (C), after correction for partial saturation of (a) when irradiating (a'), a, c, d, h, n, and x arise from ile-99 in the major form [21]; a', c', d', h', n', and x' are proposed to arise from the corresponding protons of ile-99 in the minor form.   (Reproduced from [77] with permission.)

is the spectrum of freshly prepared protoheme(IX)-reconstituted Mb-CN, chosen because the minor form [primed peaks in Fig. 19(B)] is present in greatest amount just after reconstitution. The heterogeneity is particularly clear in this figure [19(A)] with reference to the resonances a, a' which lie upfield near -9 ppm. Initially, the minor component accounts for about 45% of the total myoglobin-cyanide present. However, as time passes this minor form transforms to the "native" form, so that the difference in proton spectra between that

taken immediately after reconstitution and that taken a period of
time later results in a spectrum solely of the minor form (when
appropriate scaling is applied). This is shown in Figure 19(B).

Irradiation of resonances a' (Fig. 20) and a result in nuclear
Overhauser enhancements (negative) to peaks farther downfield. In
Figure 20 these are resonances labeled c', d', h', n', x' for irradiation of a'. For irradiation of resonance a enhancements are produced for peaks labeled c, d, h, n, x which exhibit different shift
positions from the primed peaks [83]. This connectivity pattern, the
magnitude of the Overhauser enhancements, and the transient NOE behavior allow assignment of these resonances to the isoleucine at primary
sequence position 99 (FG5) [72,84]. Thus, a protein-fixed reference
amino acid for comparing heme orientational position has been identified. The shifts and assignments of ile-99 are given in Table 5.

Consider next the relationship of ile-99 to the heme in the
two orientations achieved by rotating the heme 180° about the α-γ
heme meso axis, while maintaining the protein matrix structure constant. These two orientations are shown in Figure 21(A) and (B).
In 21(A) ile-FG5 lies close to heme methyls 3 and 5 whereas in 21(B)
it lies close to heme methyls 1 and 8. Due to this proximity NOEs

TABLE 5

ILE-99 (FG5) Chemical Shifts in Metcyanomyoglobin[a]

Assignment[b]	Major form signal[c]	[Fig. 21(A)] shift	Minor form signal[c]	[Fig. 21(B)] shift	Shift difference
γ_1-CH	a	-9.289	a'	-8.762	-0.54
δ-CH$_3$	c	-3.722	c'	-3.728	0.01
γ_2-CH$_3$	d	-3.364	d'	-3.258	-0.10
γ_1-CH'	h	-1.829	h'	-1.749	-0.08
α-CH	n	-0.380	n'	-0.276	-0.10
β-CH	x	-0.085	x'	0.024	-0.11

[a]Chemical shifts are reported in ppm relative to 2,2-dimethyl-2-
silapentane-5-sulfonate at 30°C, pH* 8.6. Digital resolution:
±0.008 ppm.
[b]From Ref. 83 for major form.
[c]Labeling as in Figure 20.
Source: Reprinted by permission from Ref. 77.

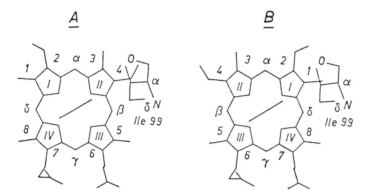

FIG. 21. (A) Orientation of ile-99 relative to the heme plane in native sperm whale myoglobin (major component) after the x-ray structure of oxymyoglobin [72]. (B) Orientation of ile-99 relative to the heme plane after rotation of the latter by 180° about its α-γ meso axis. The line pointing from pyrrole IV to II in (A), or pyrrole III to I in (B) represents the imidazole ring of the proximal histidine. (Reproduced from [77] with permission.)

should be observable from ile-99 to the 8-CH_3 or the 5-CH_3 depending on heme orientation. Note that in Figure 21(B) the heme 1-CH_3 should exhibit the largest NOE; however, this resonance is not resolved for reasons discussed above. The 1-CH_3 resonance is camoflauged by the resonances in the envelope between 0 and 10 ppm, so the 8-CH_3 resonance must be used instead. The heme methyls are then assigned by deuterium-substituted protohome(IX) as discussed above. NOE experiments are carried out by irradiating the a, a' (γ-CH on isoleucine) resonances and looking for effects in the heme methyl region of the spectrum. This is shown in Figure 22 where NOE connectivity is established between isoleucine resonance a and the heme 5-CH_3 [Fig. 22(B)] and also between isoleucine resonance a' and the heme 8-CH_3. Therefore, the major form encountered in metmyoglobin cyanide is that shown in Figure 21(A) whereas the minor form has the orientation demonstrated by 21(B).

Given data that established heme orientational isomerism as a potentially complicating but nevertheless interesting phenomenon with functional implications, it is of interest to consider what factors in heme-polypeptide interactions govern the establishment

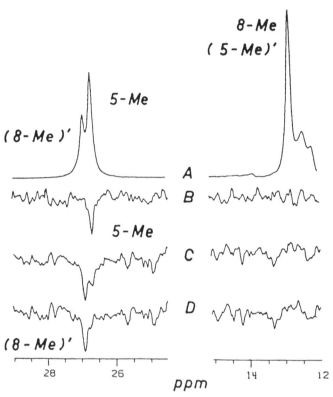

FIG. 22. The four traces correspond to the cases shown in Figure 20.
(A) Reference spectrum; only the native heme 5-methyl and 8-methyl
regions (unprimed) are shown. The minor component 5,8-methyl resonances are indicated by primes. (B) Nuclear Overhauser effect difference spectrum obtained upon saturation of (a), intensity x30.
Note the specific NOE to the major heme form 5-methyl resonance.
(C) Nuclear Overhauser difference spectrum obtained upon saturation
(a'), intensity x100. (D) Same as (C), after correction for partial
saturation of (a) when irradiating (a'). Note the specific NOE connectivity of a' with the minor form 8-methyl resonance. (Reproduced
from [77] with permission.)

of heme-centered asymmetry. Horseradish peroxidase has been recon-

stituted with several hemes that differ in their peripheral substitu-

ents. Proton NMR spectroscopy of HRP-CN has the unique capability

of determining the relative orientation of the major form which is

present (3,8-CH_3 downfield, or 1,5-CH_3 downfield) [80]. This study

revealed that there is severe steric restriction on the orientational mobility of the heme position 2 vinyl group which has been described as "steric clamping." For HRP, the heme pocket seems to have been tailored to a refined accommodation of the heme peripheral substituents on pyrroles I and II within a localized area of the heme pocket. This concept is described in greater detail in the original reference [80].

4. RESONANCES OF AXIAL LIGANDS

4.1. Assignments

As described in the first section of this chapter, the most frequently encountered axial ligands in heme proteins thus far studied by NMR are histidine and methionine. There are several approaches to assigning hyperfine shifted resonances of axial ligands including comparison with model iron porphyrin complexes [5,7,8,28], nuclear Overhauser effects [4,58-63], and isotope substitution (^{2}H for ^{1}H) for exchangeable protons [85]. For nonexchangeable protons the first two methods are extremely useful and these have been reviewed in detail elsewhere [4,5,7,8,28].

An example of the use of deuteration of exchangeable resonances is shown in Figure 6 for ferrous high-spin heme globins and this method has also been employed for the proximal histidine $N_\delta H$ assignment given in Figure 8 for high-spin ferric sperm whale myoglobin. Obtaining spectra of identical protein forms, under identical conditions, in ^2H_2O (D_2O) and 90% ^1H_2O solutions establishes the exchangeable nature of a proton of interest. Its subsequent assignment may then rely on comparison with iron porphyrin models or on a method such as measurement of T_1 as a means of gaining some geometry information that can aid in assignment. A good example of application of the relaxation method for making assignments of axial ligand resonances is given for metmyoglobin cyanide [105]. However, a more detailed recounting of this work will be deferred to Sec. 5.1 where it will be discussed in the context of making assignments to protons of amino acids which line the heme pocket.

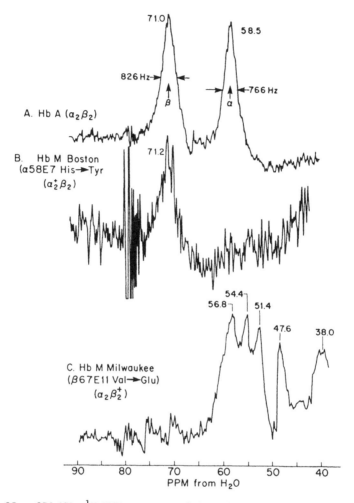

FIG. 23. 250-MHz ^{1}H NMR spectra of deoxyhemoglobins in 0.1 M bis-Tris buffer in H_2O at pH 6.7 and 27°C over the spectral region 40-90 ppm from H_2O: (A) Hb A; (B) Hb M Boston; (C) Hb M Milwaukee. The beat appearing at ~80 ppm in spectrum B is generated by the 20-kHz time-sharing operation of the spectrometer and is not a signal. (Reproduced from [88] with permission.) The results of this type of experiment yield assignments of the β- and α-proximal histidine $N_\delta H$ protons. In (B) only the β chain remains in the ferrous state and the resonance that remains visible in the $N_\delta H$ region of the spectrum can be assigned to the β-$N_\delta H$.

The most interesting case of proximal histidine resonance assignments is relative to hemoglobin function [86-92]. The tetrameric hemoglobins consist of two α and two β subunits each with similar but not identical heme pocket structures. In each case the heme iron is coordinated to a proximal histidine whose imidazole $N_\delta H$ resonance is deuterium-exchangeable, but in H_2O solutions is observable and is thus a clear indicator of the five-coordinate, high-spin ferrous deoxy form of the protein. As shown in Figures 23 and 24 and Table 6, the tetrameric hemoglobins in the ferrous deoxy forms (S = 2) exhibit two proximal histidine $N_\delta H$ resonances. As examples,

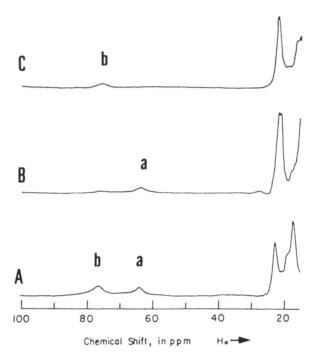

FIG. 24. Downfield portions of the 200-MHz proton NMR spectra of hemoglobin A (HbA) (A) and the valency hybrids $\alpha_2(\beta^+CN)_2$ (B) and $(\alpha^+CN)_2\beta_2$ (C) in 85% H_2O/15% ^2H_2O 0.1 M in bis-Tris, pH 6.5 at 25°C. The proximal histidyl imidazole $N_\delta H$ peaks are a (73.7 ppm) and b (76.0 ppm) in HbA, a (62.4 ppm, α subunit) in $\overline{\alpha}_2(\beta$-CN)$_2$ and b ($\overline{7}4.7$ ppm, β subunit) in $\overline{(\alpha$-CN)$_2\beta_2}$; chemical shifts are referenced to DSS. (Reproduced from [86] with permission.)

TABLE 6

Observed Shifts (in ppm) of Proximal Histidine $N_\delta H$ in
Ferrous Deoxy Subunits of Hemoglobins and Myoglobin

Protein	Temp.	pH	δ_β^a	δ_α^a	Ref.
HbA	25	6.8	75.9	63.1	85,86
	26	6.4	76.1	63.9	89
HbA-$\alpha_2(\beta CN)_2$[b]	25	6.5		62.4	86
HbA-$(\alpha CN)_2\beta_2$[b]	25	6.5	74.7		86
Hb Boston	25	6.5	73.6		86
	27	6.7	76.0		90
Hb Milwaukee	25	6.5		62.8	87,90
HbA-α chain[c]	25	6.5		77.1	87
HbA-β chain[c]	25	6.5	86.5		87
Hb rabbit[d]	25	6.8	75.9	62.2	
Sperm whale Mb	25	6.8		77.8	85

[a] Estimated accuracy in observed shifts is better than ±0.3 ppm
reported relative to DSS.
[b] Mixed oxidation state tetramers.
[c] Isolated chains.
[d] New Zealand White.

for New Zealand white rabbit these occur at 75.9 and 62.2 ppm,
whereas for human HbA they occur at 75.9 for the β subunit and 63.9
for the α subunit.

These assignments of the different resonances to α-subunit or
β-subunit histidines originate from comparing shifts with naturally
occurring methemoglobins; those in which one pair of subunits naturally exists in the ferric state. In such cases only the subunit
with a ferrous heme exhibits an exchangeable $N_\delta H$ resonance in the
region shown in Figures 23 and 24. Typical examples shown in Table 6
are Hb Boston, whose oxidized α subunit exhibits only a single resonance between 71.2 to 76.0 ppm, depending on conditions, which is
assigned to the β-subunit $N_\delta H$. Hemoglobin Milwaukee, a second human
mutant hemoglobin, but one in which the β subunit possesses a ferric

heme, displays a resonance in the region from 56.8 to 62.8 ppm
(depending on conditions and reference frequency), which is assignable to the α subunit proximal histidine. These resonances are used
to directly infer that for normal deoxy human hemoglobin A the 75.9
ppm resonance is due to the β-subunit proximal histidine $N_\delta H$ and the
63.1 ppm resonance belongs to the β-subunit $N_\delta H$. These assignments
are confirmed by creation of the mixed-valence hybrids HbA-$\alpha_2(\beta CN)_2$
and HbA-$(\alpha CN)_2\beta_2$ in which the β and α subunits, respectively, are
oxidized and cyanide-ligated (Table 6; Fig. 24).

4.2. Characterization of Quaternary State
Transitions in Hemoglobins

The proximal histidine $N_\delta H$ resonances in deoxy ferrous hemoglobins
have been used to study the hemoglobin quaternary transition. This
structural change accompanies ligand binding such as oxygenation.
Deoxyhemoglobin exists in the T, or tensed, state in which all four
of the hemes in a given molecule are in high-spin ferrous states
(S = 2). When the protein has been saturated by molecular oxygen,
all four hemes are in low-spin, six-coordinate (S = 0), oxygen-bound
forms. The quaternary structure of oxyhemoglobin is termed R, or
relaxed, and is distinct from the T state [72]. Thus, ligand binding
is associated with a T $\rightleftharpoons$ R structural transition. This summary of
quaternary structures is accurate for normal human and horse hemoglobins; however, at least one human hemoglobin mutant, Hb Kempsey,
exists in the ferrous deoxy state, not in the T quaternary structure
as expected, but rather it displays a structure best described as R
state. In the presence of the allosteric effector inositol hexa-
phosphate deoxyhemoglobin Kempsey can be switched to a T-state structure [87]. This R $\rightarrow$ T switch occurs while the heme subunits are high
spin so that the proximal histidine $N_\delta H$ is observable in both R-like
and T-like structures. As expected, the observed shifts are different
for each structure. In the R structure the $N_\delta H$ shifts are δ_α = 67.3
ppm and δ_β = 77.7 ppm whereas in the T structure δ_α = 64.6 ppm and

δ_β = 77.1 ppm for measurements at 25°C and pH = 6.5. Thus, quaternary structural changes in Hb Kempsey are manifested primarily in the α subunit, insofar as NMR is concerned [88,89].

4.3. Heme Pocket Dynamics Detected by Labile Proton Exchange

For both ferric high- and low-spin and ferrous high-spin heme proteins, the proximal histidine's imidazole $N_\delta H$ provides a unique and specific reporter for dynamic fluctuations in the heme pocket. For several proteins the dynamics of proximal histidine $N_\delta H$ stability have been studied [93-95]. Hemoglobin A has been the object of much tritium isotope exchange work [96,97] which has had as its objective the delineation of rapid structural fluctuations in the protein which calculations [73,74] indicate must occur in the course of ligand binding.

This problem has been approached from a more specific point of view utilizing the proximal histidine $N_\delta H$ resonances of human HbA. As indicated in Sec. 4.2, these resonances occur near 76 and 63 ppm for the β and α subunits, respectively [93]. The procedure that is followed for T-state deoxyhemoglobin (S = 2) is to follow the disappearance of the labile histidine ring proton resonances following dilution of a 90% ^1H_2O hemoglobin solution with ^2H_2O. Resonance intensity is lost as time passes due to ^1H for ^2H exchange of the $N_\delta H$ as shown in Figure 25 (left side) [93]. Similar experiments can be carried out for oxyhemoglobin A (Figure 25, right side) although the technique is slightly more complicated. For HbO_2 the sample must be converted to the high-spin deoxy form by treatment of HbO_2 samples with dithionite ion at various times after the deuterium oxide dilution in order to quantitate the extent of deuterium exchange using the proximal $N_\delta H$. The results indicate that in the T state (deoxy form) the α subunit exhibits an exchange half-life of approximately 1.5×10^4 sec, while the β-subunit proximal $N_\delta H$ exchanges more slowly with a demonstrated half-life of about 2.6×10^5 sec at 25°C. Results

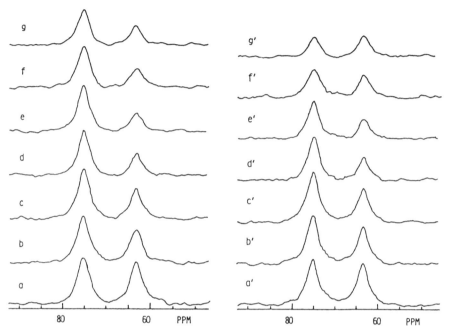

FIG. 25. The time evolution of the ^{1}H-^{2}H exchange of the proximal histidyl NH resonances, α (64 ppm) and β (76 ppm) subunits of HbA after mixing deoxy HbA in H_2O with ^2H_2O to give a final ^2H_2O concentration of 50% is shown in traces a-g. Deoxy HbA was in 0.1 M bis-Tris, 0.2 M NaCl, at pH 6.97 and 25°C. The traces correspond to the time after mixing with ^2H_2O: (a) 7 min, (b) 9 x 10^1 min, (c) 1.5 x 10^2 min, (d) 2.5 x 10^2 min, (e) 4.2 x 10^2 min, (f) 9.0 x 10^2 min, (g) 1.7 x 10^3 min. Traces a'-g' reflect analogous time evolution with HbO$_2$. ^2H_2O was added to HbO$_2$ to give a final ^2H_2O concentration of ~55%. After a variable mixing time in which ^{1}H-^{2}H exchanged in HbO$_2$, we introduced dithionite to remove the O$_2$ and monitored the extent of exchange with the intensity loss of the proximal histidyl NH resonances in the deoxy form of HbA. HbO$_2$ was also in 0.1 M bis-Tris 0.2 M NaCl at pH 6.95 and 25°C. Traces a'-g' reflect the ^{1}H-^{2}H mixing time in HbO$_2$: (a') 2 x 10^{-1} min, (b') 6 x 10^{-1} min, (c') 1.5 min, (d') 2.5 min, (e') 3 x 10^1 min, (f') 1.2 x 10^2 min, (g') 1.8 x 10^2 min. (Reproduced from [93] with permission.)

for the R state (oxyhemoglobin) indicate that the half-lives for exchange of the α and β subunits are, respectively, nearly 40 and 80 times less than for the T-state case. The increased $N_\delta H$ exchange rate observed in the R state is apparently related to the difference

in quaternary structures and certainly suggests greater solvent-heme
crevice communication in the R-state hemoglobin.

5. RESONANCES OF THE HEME POCKET NOT ATTRIBUTABLE TO HEME LIGANDS

5.1. Detection

If a selection of specifically deuterated hemes is available such
that all possible hyperfine resonances originating from the heme
group protons are assigned, one may find a number of other resonances that do not belong to the heme but appear in the hyperfine
shift region. Resonances belonging to axial ligands (histidine,
methionine) may be assigned by comparison to iron porphyrin models
[5,7,28]. Frequently, in low-spin ferric and high-spin ferrous heme
proteins, resonances unassigned by these methods occur in the hyperfine shift region. These must result from amino acids that lie in
proximity to the heme but are not directly bonded to the iron ion.
Their shifts are, in general, a consequence of the paramagnetic
dipolar shift, ring current shift, and hydrogen bonding. Frequently,
these resonances are not present in deuterium oxide solutions, indicating that they are labile to isotope exchange and consequently
originate with protons that are O- or N-bonded. For tetrameric
hemoglobins several isotope-exchangeable and nonexchangeable hyperfine resonances have been identified for the ferrous high-spin forms
in the region 70 to 10 ppm downfield [98-100]. This same region is
featureless for the low-spin, diamagnetic hemoglobin form, HbCO.
This comparison between diamagnetic and paramagnetic forms is therefore one method for elucidating the origin of such resonances.

For isotope-exchangeable resonances comparison between otherwise identical protein solutions in ^2H_2O vs. 90% ^1H_2O identify potential hyperfine shifted resonances just as for the proximal histidine
$N_\delta H$ discussed previously. For ferric low-spin proteins CcP-CN is a
typical example. Figure 26(A) shows the CcP-CN proton spectrum in
90% ^1H_2O whereas Figure 26(B) shows an identical preparation in 99.8%

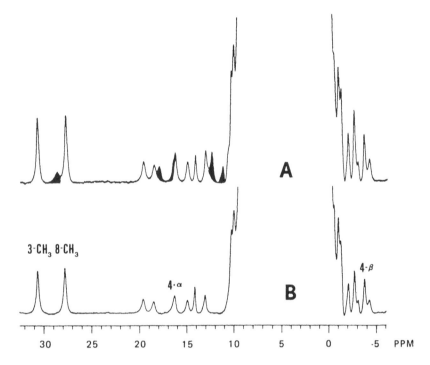

FIG. 26. Proton NMR spectra of cytochrome c peroxidase-cyanide
solutions. Each solution was 2.4 mM in CcP, 0.1 M KNO_3, and spectra
were accumulated at 23°C. (A) A 90% ^1H_2O solution, pH = 7.3; (B) A
99.8% ^2H_2O solution that gave a pH meter reading of 6.9.

^2H_2O. It is obvious from a comparison of the two spectra that several isotope-exchangeable resonances with significant hyperfine
shifts are present in Figure 26(A). These resonances are shown as
shaded areas and must originate from protons of amino acid side
chains in the heme pocket that lie close enough to the heme to be
affected by the through-space dipolar shift term.

Confirmation of the origin of such shifts makes use of the
fact that the hyperfine shift equations [Eqs. (6)-(8)] indicate that
hyperfine resonances are predicted to demonstrate temperature dependence. Accordingly, all of the hyperfine resonances in the downfield and upfield regions of Figure 26 have been shown to exhibit
temperature-dependent shifts [101,102].

5.2. Assignment

Assigning hyperfine proton resonances that are not attributable to
the heme group, or its ligands, to specific amino acids is a significant task. Comparing spectra from a homologous series of proteins,
say human mutant hemoglobins, in which the primary sequence deletions
and substitutions are known is one method that has been used [98-100,
103]. Another potential method is the chemical modification or photo-
chemical destruction of accessible amino acids coupled with comparison
of spectra between the holoprotein and the modified protein. A more
sophisticated but similar approach involves the technique of site-
directed mutagenesis in which primary sequence changes in a protein
are established using recombinant DNA methods. Recent efforts to
produce mutants of cytochrome c peroxidase and cytochrome c indicate
that this technique will soon become the method of choice for protein
NMR laboratories.

Another method that seems reliable, if not broadly applied, is
based on relative dipolar relaxation rates [104,105]. It is possible
to show that under certain conditions the ratio of dipolar relaxation
rates for two protons A and B at respective distances r_A and r_B from
the iron center of the heme is given approximately by Eq. (12) [104,
105]. This equation is derived from Eq. (9) which takes into account

$$\frac{T_1^A}{T_1^B} = \frac{r_A^6}{r_B^6} \tag{12}$$

only dipolar relaxation from electrons that are centered on the iron
ion. Despite what has been an apparently successful application of
this method for assigning the proximal and distal histidine $N_\delta H$ protons in Mb-CN [105], these same authors have recently pointed out
that use of the differential dipolar relaxation method without cor-
recting for relaxation induced by the electron magnetic moment that
is ligand-centered is inherently incorrect [104]. Thus, for low-spin
ferric heme proteins an equation such as (12) is expected to apply to
protons of amino acids not directly coordinated to the heme iron,

provided they are in close proximity to the heme. However, due to metal-ligand covalency the dipolar relaxation equation (9) actually consists of two terms, the metal-centered term given in Eq. (9) and a term due to the delocalized spin density (ligand-centered).

6. PROTON NMR STUDIES OF PROTEIN-PROTEIN COMPLEXES

As a final illustration of the value of hyperfine proton resonances in heme protein research I would like to indicate some recent work occurring in our laboratory, as well as others, directed at the so-called molecular docking complexes involving heme proteins. Molecular complexes such as these have been regarded as necessary in biochemical pathways involving electron transfer. These include photosynthetic, mitochondrial electron transport, respiratory, and microsomal electron transfer pathways.

Perhaps the most detailed picture of how heme proteins might form bimolecular complexes comes from the work of Poulos and Kraut [106,107] which is based on their crystal structure of cytochrome c peroxidase [108,109] and structures of the cytochromes c [26,110-113]. This work has been summarized by Kraut [114]. The technique involves developing a structure for a complex between two heme proteins based on the three-dimensional alignment of complementary acidic and basic surface amino acids on each protein. For CcP and cytochrome c this involves a ring of lysines that appear distributed about the cytochrome c surface, surrounding the accessible heme edge, and a set of aspartate side chains on the surface of cytochrome c peroxidase. When the CcP aspartates (primary sequence positions 37, 79, 216) and the cytochrome c lysines (primary sequence positions 13, 27, 72) are juxtaposed, a three-dimensional structure with the heme planes of the two proteins parallel is achieved. This is shown schematically in Figure 27. This hypothetical structure has been the source for construction of a model for electron transfer between these two proteins [106,107,114].

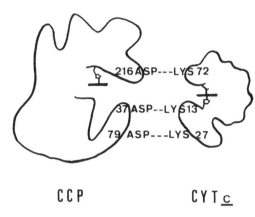

CCP   CYT c

FIG. 27. A schematic representation of the complex formed between
cytochrome c peroxidase (left) and cytochrome c (right). This drawing is based on the Poulos-Kraut [106,107] model and indicates the
interaction between negatively charged aspartate residues on CcP and
lysines on cytochrome c which stabilizes the complex.

Such a model is physiologically relevant for yeast cytochrome c
peroxidase because the protein exhibits a very specific peroxidase
function, catalysis of the hydrogen peroxide oxidation of reduced
(ferrous) cytochrome c. CcP shows no significant reaction rate for
most of the chemical reactions that other heme peroxidases (chloroperoxidase, horseradish peroxidase) engage in. The CcP function is
summarized by the reaction sequence given in Eqs. (13)-(15), where

$$CcP + H_2O_2 \rightarrow CcP\text{-}I \tag{13}$$

$$CcP\text{-}I + Cyt\ c\ (2+) \rightarrow CcP\text{-}II + Cyt\ c\ (3+) \tag{14}$$

$$CcP\text{-}II + Cyt\ c\ (2+) \rightarrow CcP + Cyt\ c\ (3+) \tag{15}$$

CcP-I and CcP-II are the ferryl oxidized intermediates, referred to
as CcP compound I and CcP compound II. The electron transfers between
oxidized CcP and reduced cytochrome c [Eqs. (14) and (15)] are pre-
sumed to require specific complex formation between these two proteins. This is the complex that Poulos and Kraut have modeled and
which we are attempting to study in solution.

Other models with differing degrees of biological relevance
have also been constructed by model-building methods, including

cytochrome b with cytochrome c [115]; cytochrome c with methemoglobin
[116]; cytochrome c with flavodoxin [117,118] and cytochrome c with
metmyoglobin [119]. Furthermore, protein-protein complexes have been
studied by proton NMR for cytochrome c peroxidase with cytochrome c
[120-122]; for plastocyanin with cytochrome c [123]; for cytochrome
b_5 with cytochrome c [124]; and for cytochrome c with cytochrome c
oxidase [125].

Our work with the CcP/cytochrome c complex involves formation
of both the noncovalent complex and the complex formed by cross-
linking the two proteins employing a carbodiimide [126,127]. The
early data on the noncovalent complex indicated a stoichiometry of
approximately 1:1 (CcP/cytochrome c) and a fully complexed linewidth
of 100 Hz, although these data were based on a four-point graph [120],
making them somewhat unreliable. The data from our work [121] shown
in Figures 28 and 29 provide enough points to definitely assign the
stoichiometry as 1:1. Figure 28 shows the behavior of the ferric
cytochrome c heme 8-CH_3 and 3-CH_3 hyperfine shifts as a function of
the cytochrome c/cytochrome c peroxidase mole ratio. In this experiment the pH was rigorously maintained (meter reading of 6.8 ± 0.2
in 99% ^2H_2O) and the total protein concentration was kept constant
at 2.8 mM, in contrast to the previous work [120]. The shifts change
smoothly until the 1:1 mole ratio is obtained, whereupon no further
shift is induced (Fig. 29). The cytochrome c methyl linewidths also
increase as the complex is formed, reflecting the increased effective
molecular weight of cytochrome c in the complex and the associated
longer rotational correlation time. Figure 30 shows the complete
proton spectrum of the 1:1 complex with assigned resonances to both
the CcP and cytochrome c protons. Integration of the relative methyl
resonance areas for CcP and cytochrome c indicate that all normally
visible methyl protons are detected in the complex.

We have pursued studies of the covalently crosslinked complex
between cytochrome c peroxidase and cytochromes c from several species. Shown in Figure 31 is the full proton spectrum of the covalent
complex formed between CcP and horse ferricytochrome c. The cross-
linking technique employed for these types of studies is a literature

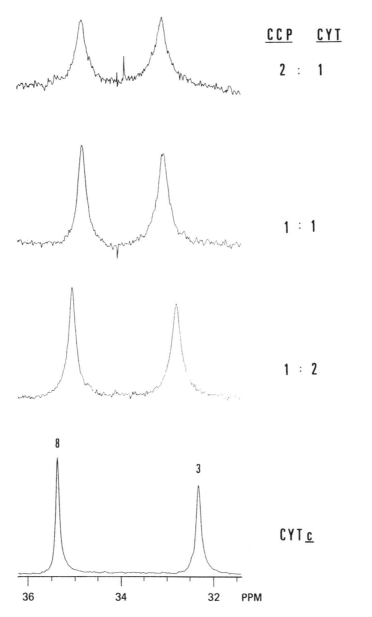

FIG. 28. Proton NMR spectra of the cytochrome c heme methyl resonances (8-CH$_3$, 3-CH$_3$) as a function of changing the mole ratio of
cytochrome c peroxidase to cytochrome c. This figure shows the
linewidth and shift changes that occur as a consequence of complex
formation between the two proteins. These spectra were recorded
from solutions maintained at constant total protein concentration,
at 22°C, in 10 mM KNO$_3$/^{2}H$_2$O solution, with a pH meter reading of 6.8.

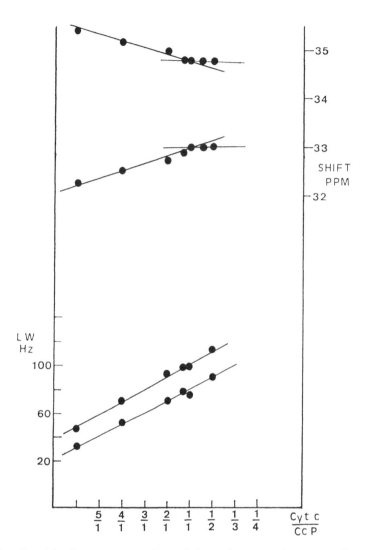

FIG. 29. Graphical representation of data for experiments such as those described in the Figure 28 caption. The upper part of this figure shows that the cytochrome c methyl resonance positions (8-CH$_3$ upper trace, 3-CH$_3$ lower trace) change smoothly until the peroxidase/cytochrome mole ratio is 1:1. This indicates how the cytochrome 3-CH$_3$ (upper line) and 8-CH$_3$ (lower line) linewidths vary with the mole ratio of peroxidase to cytochrome.

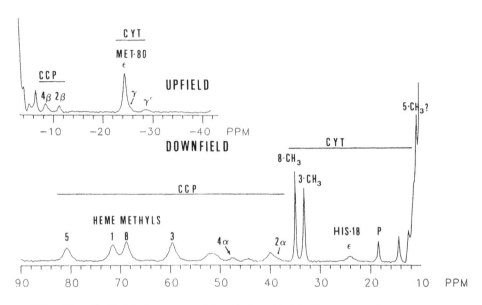

FIG. 30. Upfield (upper part) and downfield (lower part) proton
hyperfine shift regions of the cytochrome c peroxidase/cytochrome c
complex. Assignments are taken from the assignments for the individual proton resonances described in the text.

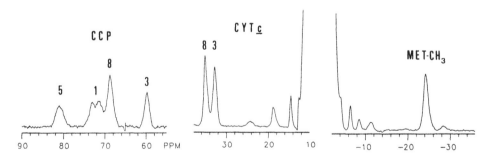

FIG. 31. A 361-MHz proton NMR spectrum of the covalently crosslinked
complex formed between cytochrome c peroxidase and cytochrome c. CcP
and cytochrome c resonance positions are similar to those shown in
Figure 30 for the noncovalent complex. Assignments indicated in this
figure come from assignments for the individual proteins.

method [127], although we have improved the isolation method in order to purify milligram quantities of the complex for NMR studies [126]. Of interest are the obvious splitting of the CcP 1-CH_3 resonance, the fact that the cytochrome c heme methyl linewidths (at half resonance height) are nearly 300 Hz, and the direct integrations of CcP and cytochrome c methyl resonances that unambiguously demonstrate the 1:1 stoichiometry of this covalent complex. As far as these spectra show there is no contradiction to the proposed molecular docking complex of these two proteins that is visualized by Poulos and Kraut [106,107].

7. CONCLUSION

This chapter was meant to introduce some of the fundamental principles and practical considerations currently applied in proton NMR studies of paramagnetic heme proteins. Those of us interested in studying these creations of nature cannot help but be impressed by their structural and functional diversity. Professor Max Perutz illustrated this beautifully when he related how Keilin would ask of him "how nature could use the same haem for so many different functions, simply by attaching it to different proteins" [128]. It will become clear to those perusing the literature in this area that NMR cannot stand by itself in the study of heme proteins. In the same sentiment of diversity that Keilin's remark to Perutz embodies, realization of the full potential of NMR applications frequently requires that information from many areas (kinetics, crystallography, other spectroscopies, etc.) be combined. The examples presented in this chapter uniquely illustrate how proton NMR can often resolve questions (or create new ones) that other methods have identified.

I have not attempted a comprehensive review of this area but rather chose to illustrate the highlights of current research efforts. I have drawn data for many of the figures from spectra that are in my own files, both published and unpublished, as well as from the work of others. I have neglected several areas including the use of

coordinated ligands, carbon and deuterium nuclei, and solid state
applications. All of these areas have been the subject of published
work in 1983-1985. Some of these are described in a recently completed review [8]. Nevertheless, it is hoped that this work illustrates the wealth of information that proton NMR studies of paramagnetic heme proteins are capable of developing.

ACKNOWLEDGMENTS

I would specifically like to thank Professors Gerd LaMar, Kurt
Wüthrich, and Chien Ho for giving me their permission to use figures
from their published work in this article. The unpublished work
from my laboratory cited here has been supported by the National
Institutes of Health (AM30912) and the National Science Foundation
(PCM-DMB 8403353). Finally, I wish to thank the Alfred P. Sloan
Foundation for a Fellowship whose duration encompassed the prepara-
tion of this manuscript.

ABBREVIATIONS

CcP	cytochrome c peroxidase
dss	2,2-dimethyl-2-silapentane-5-sulfonate
Hb	hemoglobin
HRP	horseradish peroxidase
Mb	myoglobin
NOE	nuclear Overhauser effect
tms	tetramethylsilane

REFERENCES

1. D. Dolphin (ed.), *The Porphyrins*, Vols. 1-7, Academic Press, New York, 1978, 1979.

2. K. Smith (ed.), *Porphyrins and Metalloporphyrins*, Elsevier, Amsterdam, 1975.

3. R. Bonnett, in *The Porphyrins* (D. Dolphin, ed.), Academic Press, New York, 1978, Vol. 1, pp. 1-27.

4. R. M. Keller and K. Wüthrich, in *Biological Magnetic Resonance* (L. J. Berliner and J. Reuben, eds.), Plenum Press, New York, Vol. 3, 1981.

5. G. N. LaMar, in *Biological Applications of Magnetic Resonance* (R. G. Shulman, ed.), Academic Press, New York, 1979, pp. 305-343.

6. C. Ho and I. M. Russu, *Meth. Enzymol., 76,* 275-312 (1981).

7. H. Goff, in *Iron Prophyrins* (A. B. P. Lever and H. B. Gray, eds.), Addison-Wesley, Reading, Mass., 1983, Pt. 1, pp. 239-281.

8. J. D. Satterlee, *Ann. Repts. NMR Spectrosc., 17,* 79-179 (1986).

9. R. Lemberg and J. E. Falk, *Biochem. J., 49,* 674 (1951).

10. R. K. DiNello and C. K. Chang, in *The Porphyrins* (D. Dolphin, ed.), Academic Press, New York, 1978, Vol. 1, pp. 289-339.

11. M. R. Ondrias, S. D. Carson, M. Hirasawa, and D. B. Knaff, *Biochim. Biophys. Acta, 830,* 159-163 (1985).

12. R. Lemberg and J. Barrett, *Cytochromes*, Academic Press, London, 1973, pp. 233-245.

13. G. S. Jacob and W. H. Orme-Johnson, *Biochemistry, 18,* 2967-2975 (1979).

14. B. A. Wittenberg, L. Kampa, J. B. Wittenberg, W. E. Blumberg, and J. Peisach, *J. Biol. Chem., 243,* 1863-1878 (1968).

15. W. E. Blumberg, J. Peisach, B. A. Wittenberg, and J. B. Wittenberg, *J. Biol. Chem., 243,* 1854-1862 (1968).

16. J. A. Peterson and B. W. Griffin, *Arch. Biochem. Biophys., 151,* 427-433 (1972).

17. B. W. Griffin, J. A. Peterson, and R. W. Estabrook, in *The Porphyrins* (D. Dolphin, ed.), Academic Press, New York, 1978, Vol. 7, pp. 333-376.

18. P. F. Hollenberg and L. P. Hager, *J. Biol. Chem., 248,* 2630-2635 (1973).

19. T. Odajima and I. Yamazaki, *Biochem. Biophys. Acta, 206,* 71-79 (1970).

20. D. Keilin and T. Mann, *Proc. R. Soc. London Ser. B, 122,* 119 (1937).

21. H. A. Harbury, *J. Biol. Chem., 225,* 1009-1024 (1957).

22. I. Yamazaki and L. H. Piette, *Biochim. Biophys. Acta., 77,* 47-64 (1963).

23. J. B. Wittenberg, R. W. Noble, B. A. Wittenberg, E. Antonini, M. Brunori, and J. Wyman, *J. Biol. Chem., 242,* 626-634 (1967).

24. J. A. Peterson, Y. Ishimura, and B. W. Griffin, *Arch. Biochem. Biophys., 149,* 197-208 (1972).

25. P. Argos and F. S. Mathews, *J. Biol. Chem., 250,* 747-751 (1975).

26. T. Takano, B. L. Trus, N. Mandel, G. Mandel, O. B. Kallai, R. Swanson, and R. E. Dickerson, *J. Biol. Chem., 252,* 776-785 (1977).

27. T. J. Swift, in *NMR of Paramagnetic Molecules* (G. N. LaMar, W. deW. Horrocks, and R. H. Holm, eds.), Academic Press, New York, 1973, pp. 53-83.

28. G. N. LaMar and F. A. Walker, in *The Porphyrins* (D. Dolphin, ed.), Academic Press, New York, Vol. 7, 1978.

29. J. P. Jesson, in *NMR in Paramagnetic Molecules* (G. N. LaMar, W. deW. Horrocks, and R. H. Holm, eds.), Academic Press, New York, 1973, pp. 1-52.

30. O. Jardetzky and G. C. K. Roberts, *NMR in Molecular Biology,* Academic Press, New York, 1981.

31. R. M. Keller, O. Groundinsky, and K. Wüthrich, *Biochim. Biophys. Acta., 427,* 497-511 (1976).

32. J. D. Satterlee, G. N. LaMar, and T. J. Bold, *J. Am. Chem. Soc., 99,* 1088-1096 (1977).

33. W. deW. Horrocks, in *NMR of Paramagnetic Molecules* (G. N. LaMar, W. deW. Horrocks, and R. H. Holm, eds.), Academic Press, New York, 1973, pp. 127-177.

34. I. Solomon, *Phys. Rev., 99,* 559-565 (1955).

35. I. Solomon and N. Bloembergen, *J. Chem. Phys., 25,* 261-266 (1956).

36. J. Kowalewski, L. Nordenskiöld, N. Benetis, and P. O. Westlund, *Progress NMR Spectrosc., 17,* 141-185 (1985).

37. M. Gueron, *J. Mag. Reson., 19,* 58-66 (1975).

38. A. J. Vega and D. Fiat, *Mol. Phys., 31,* 347-355 (1976).

39. K. Wüthrich, J. Hochman, R. M. Keller, G. Wagner, M. Brunori, and G. Giacometti, *J. Mag. Reson., 19,* 111-113 (1975).

40. M. E. Johnson, L. W. M. Fung, and C. Ho, *J. Am. Chem. Soc., 99,* 1245-1250 (1977).

41. G. N. LaMar, J. T. Jackson, and R. G. Bartsch, *J. Am. Chem. Soc.*, *103*, 4405-4410 (1981).

42. I. Morishima and S. Ogawa, *Biochemistry*, *17*, 4384-4388 (1978).

43. G. N. LaMar, V. P. Chacko, and J. S. deRopp, in *The Biological Chemistry of Iron* (H. B. Dunford, D. Dolphin, K. Raymond, and L. Sieker, eds.), Reidel, New York, 1982, pp. 357-373.

44. J. E. Erman and J. D. Satterlee, in *Electron Transport and Oxygen Utilization* (C. Ho, ed.), Elsevier North Holland, Amsterdam, 1982, pp. 223-227.

45. J. D. Satterlee and J. E. Erman, *J. Biol. Chem.*, *256*, 1091-1093 (1981).

46. G. N. LaMar, J. S. deRopp, L. Latos-Grazynski, A. L. Balch, R. B. Johnson, K. M. Smith, D. W. Parish, and R. J. Cheng, *J. Am. Chem. Soc.*, *105*, 782-787 (1983).

47. J. D. Satterlee, *Inorg. Chim. Acta*, *79*, 195-196 (1983).

48. K. M. Smith, *Acct. Chem. Res.*, *12*, 374-381 (1979).

49. K. M. Smith, E. M. Fuginari, K. C. Langry, D. W. Parish, and H. D. Tabba, *J. Am. Chem. Soc.*, *105*, 6638-6646 (1983).

50. J. A. S. Cavaleiro, A. M. D'A. Rocha Gonsalves, G. W. Kenner, K. M. Smith, R. G. Shulman, A. Mayer, and T. Yamane, *J. Mol. Biol.*, *86*, 749-756 (1974).

51. B. Evans, K. M. Smith, G. N. LaMar, and D. B. Viscio, *J. Am. Chem. Soc.*, *99*, 7070-7072 (1977).

52. G. N. LaMar, J. S. deRopp, K. M. Smith, and K. C. Langry, *J. Biol. Chem.*, *255*, 6646-6652 (1980).

53. G. N. LaMar, J. S. deRopp, K. M. Smith, and K. C. Langry, *J. Am. Chem. Soc.*, *102*, 4833-4835 (1980).

54. J. D. Satterlee, J. E. Erman, G. N. LaMar, K. M. Smith, and K. C. Langry, *J. Am. Chem. Soc.*, *105*, 2099-2104 (1983).

55. J. D. Satterlee, J. E. Erman, G. N. LaMar, K. M. Smith, and K. C. Langry, *Biochim. Biophys. Acta.*, *743*, 246-255 (1983).

56. G. N. LaMar, P. D. Burns, J. T. Jackson, K. M. Smith, K. C. Langry, and P. Strittmatter, *J. Biol. Chem.*, *256*, 6075-6079 (1981).

57. J. D. Satterlee and J. E. Erman, *J. Am. Chem. Soc.*, *103*, 199-200 (1981).

58. G. Wagner and K. Wüthrich, *J. Magn. Reson.*, *33*, 675-680 (1979).

59. R. M. Keller and K. Wüthrich, *Biochim. Biophys. Acta.*, *621*, 204-217 (1980).

60. H. Senn and K. Wüthrich, *Biochim. Biophys. Acta.*, *746*, 48-60 (1983).

61. H. Senn, M. Cusanovich, and K. Wüthrich, *Biochem. Biophys. Acta., 785*, 46-53 (1984).

62. R. M. Keller and K. Wüthrich, *Biochem. Biophys. Res. Commun., 83*, 1132-1139 (1978).

63. H. Senn, R. M. Keller, and K. Wüthrich, *Biochem. Biophys. Res. Commun., 92*, 1362-1369 (1980).

64. S. W. Unger, J. T. J. LeComte, and G. N. LaMar, *J. Mag. Reson., 64*, 521-526 (1985).

65. R. G. Shulman, S. H. Glarum, and M. Karplus, *J. Mol. Biol., 57*, 93-115 (1971).

66. G. N. LaMar, D. B. Viscio, K. M. Smith, W. S. Caughey, and M. L. Smith, *J. Am. Chem. Soc., 100*, 8085-8092 (1978).

67. F. A. Walker, *J. Am. Chem. Soc., 102*, 3254-3256 (1980).

68. F. A. Walker, J. Buehler, J. T. West, and J. L. Hinds, *J. Am. Chem. Soc., 105*, 6923-6929 (1983).

69. R. M. Keller and K. Wüthrich, *Biochim. Biophys. Acta., 533*, 195-208 (1978).

70. G. N. LaMar, T. L. Bold, and J. D. Satterlee, *Biochim. Biophys. Acta., 498*, 189-197 (1978).

71. T. G. Traylor and A. P. Berzinis, *J. Am. Chem. Soc., 102*, 2844-2846 (1980).

72. G. Fermi and M. F. Perutz, *Atlas of Molecular Structures in Biology, Vol. 2, Haemoglobin and Myoglobin* (D. C. Phillips and F. M. Richards, eds.), Oxford University Press, Oxford, 1981.

73. D. A. Case and M. Karplus, *J. Mol. Biol., 132*, 343-346 (1979).

74. J. A. McCammon and M. Karplus, *Accounts Chem. Res., 16*, 187-193 (1983).

75. G. N. LaMar, N. L. Davis, D. W. Parish, and K. M. Smith, *J. Mol. Biol., 168*, 887-896 (1983).

76. T. Jue, R. Krishnamoorthi, and G. N. LaMar, *J. Am. Chem. Soc., 105*, 5701-5703 (1983).

77. J. T. J. LeComte, R. D. Johnson, and G. N. LaMar, *Biochim. Biophys. Acta., 829*, 268-274 (1985).

78. G. N. LaMar, R. R. Anderson, V. P. Chacko, and K. Gersonde, *Eur. J. Biochem., 136*, 161-166 (1983).

79. G. N. LaMar, K. M. Smith, K. Gersonde, H. Sick, and M. Overkamp, *J. Biol. Chem., 255*, 66-70 (1980).

80. G. N. LaMar, J. S. deRopp, K. M. Smith, and K. C. Langry, *J. Am. Chem. Soc., 105*, 4576-4580 (1983).

81. G. N. LaMar, M. Overkamp, H. Sick, and K. Gersonde, *Biochemistry, 17*, 352-361 (1978).

82. D. J. Livingston, N. L. Davis, G. N. LaMar, and W. D. Brown, *J. Am. Chem. Soc., 106,* 3025-3026 (1984).

83. S. Ramaprasad, R. D. Johnson, and G. N. LaMar, *J. Am. Chem. Soc., 106,* 5330-5335 (1984).

84. S. E. V. Phillips, *J. Mol. Biol., 42,* 531-554 (1980).

85. G. N. LaMar, D. L. Budd, and H. Goff, *Biochem. Biophys. Res. Commun., 77,* 104-110 (1977).

86. G. N. LaMar, K. Nagai, T. Jue, D. L. Budd, K. Gersonde, H. Sick, T. Kagimoto, A. Hayashi, and F. Taketa, *Biochem. Biophys. Res. Commun., 96,* 1172-1177 (1980).

87. K. Nagai, G. N. LaMar, T. Jue, and H. F. Bunn, *Biochemistry, 21,* 842-847 (1982).

88. S. Takahashi, A. K. L. C. Lin, and C. Ho, *Biophys. J., 39,* 33-40 (1982).

89. S. Miura and C. Ho, *Biochemistry, 23,* 2492-2499 (1984).

90. S. Takahashi, A. K. L. C. Lin, and C. Ho, *Biochemistry, 19,* 5196-5202 (1980).

91. C. Dalvit, S. Miura, A. DeYoung, R. W. Noble, M. Cerdonio, and C. Ho, *Eur. J. Biochem., 141,* 255-259 (1984).

92. G. N. LaMar, T. Jue, B. M. Hoffman, and K. Nagai, *J. Mol. Biol., 178,* 929-939 (1984).

93. T. Jue, G. N. LaMar, K. Han, and Y. Yamamoto, *Biophys. J., 46,* 117-120 (1984).

94. G. N. LaMar, J. D. Cutnell, and S. B. Kong, *Biophys. J., 34,* 217-226 (1981).

95. S. B. Kong, J. D. Cutnell, and G. N. LaMar, *J. Biol. Chem., 258,* 3843-3849 (1983).

96. C. K. Woodward and B. D. Hilton, *Annu. Rev. Biophys. Bioeng., 8,* 991-1027 (1979).

97. S. W. Englander, D. B. Calhoun, J. J. Englander, R. K. Kallenbach, H. Lie, E. Malin, C. Mandal, and J. R. Rogers, *Biophys. J., 32,* 577-590 (1980).

98. C. Ho, C. J. Lam, S. Takahashi, and G. Viggiano, in *Hemoglobin and Oxygen Binding* (C. Ho, ed.), Elsevier North Holland, Amsterdam, 1982, pp. 141-149.

99. C. Ho and I. M. Russu, *Meth. Enzymol., 76,* 275-312 (1981).

100. S. Miura and C. Ho, *Biochemistry, 24,* 6280-6287 (1982).

101. J. D. Satterlee and J. E. Erman, *J. Biol. Chem., 258,* 1050-1056 (1983).

102. J. D. Satterlee and J. E. Erman, unpublished data.

103. K. J. Wiechelman, J. Fox, P. R. McCurdy, and C. Ho, *Biochemistry, 17,* 791-795 (1978).

104. S. W. Unger, T. Jue, and G. N. LaMar, *J. Mag. Reson., 61,* 448-456 (1985).

105. J. D. Cutnell, G. N. LaMar, and S. B. Kong, *J. Am. Chem. Soc., 103,* 3567-3572 (1981).

106. T. L. Poulos and J. Kraut, *J. Biol. Chem., 255,* 8199-8205 (1980).

107. T. L. Poulos and J. Kraut, *J. Biol. Chem., 255,* 10322-10330 (1980).

108. T. L. Poulos, S. T. Freer, R. A. Alden, N. H. Xuong, S. L. Edwardo, R. C. Hamlin, and J. Kraut, *J. Biol. Chem., 253,* 3730-3735 (1978).

109. T. L. Poulos, S. T. Freer, R. A. Alden, S. L. Edwardo, U. Skoglund, K. Takio, B. Eriksson, N. H. Xuong, T. Yonetani, and J. Kraut, *J. Biol. Chem., 255,* 575-580 (1980).

110. F. R. Salemme, *Ann. Rev. Biochem., 46,* 299-329 (1977).

111. R. E. Dickerson, *Sci. Am., 242,* 136-153 (1980).

112. R. E. Dickerson, T. Takano, D. Eisenberg, O. B. Kallai, L. Samson, A. Cooper, and E. Margoliash, *J. Biol. Chem., 246,* 1511-1535 (1971).

113. R. Timkovich, in *The Porphyrins* (D. Dolphin, ed.), Academic Press, New York, Vol. 7B, pp. 241-294.

114. J. Kraut, *Biochem. Soc. Trans., 9,* 197-202 (1981).

115. F. R. Salemme, *J. Mol. Biol., 102,* 563-568 (1976).

116. T. L. Poulos and A. G. Mauk, *J. Biol. Chem., 258,* 7369-7373 (1983).

117. R. P. Simondsen, P. C. Weber, F. R. Salemme, and G. Tollin, *Biochemistry, 21,* 6366-6375 (1982).

118. J. B. Matthew, P. C. Weber, F. R. Salemme, and F. M. Richards, *Nature (London), 301,* 169-171 (1983).

119. J. D. Satterlee, unpublished results (1985).

120. R. Gupta and T. Yonetani, *Biochim. Biophys. Acta, 292,* 502-508 (1973).

121. S. Moench, J. D. Satterlee, and J. E. Erman, unpublished results (1985).

122. A. P. Boswell, G. J. McClune, G. R. Moore, R. J. P. Williams, G. W. Pettigrew, T. Inubushi, T. Yonetani, and D. E. Harris, *Biochem. Soc. Trans., 8,* 637-638 (1980).

123. G. C. King, R. A. Binstead, and P. E. Wright, *Biochim. Biophys. Acta., 806,* 262-271 (1985).

124. C. G. S. Eley and G. R. Moore, *Biochem. J.*, *215*, 11-21 (1983).

125. K. E. Falk and J. Angström, *Biochim. Biophys. Acta.*, *722*, 291-296 (1983).

126. S. Moench, J. D. Satterlee, and J. E. Erman, *J. Biol. Chem.* (1986), submitted.

127. B. Waldmeyer and H. R. Bosshard, *J. Biol. Chem.*, *260*, 5184-5190 (1985).

128. M. Perutz, *Ann. Rev. Biochem.*, *48*, 328 (1979).

129. H. Senn, A. Eugster, and K. Wüthrich, *Biochim. Biophys. Acta.*, *743*, 58-68 (1983).

130. R. M. Keller and K. Wüthrich, *Biochim. Biophys. Acta*, *533*, 195-208 (1978).

131. C. C. McDonald and W. D. Phillips, *Biochemistry*, *12*, 3170-3186 (1973).

132. G. N. LaMar, J. S. deRopp, V. P. Chacko, J. D. Satterlee, and J. E. Erman, *Biochim. Biophys. Acta.*, *108*, 317-325 (1982).

133. R. Timkovich, M. S. Cork, R. B. Gennis, and P. Y. Johnson, *J. Am. Chem. Soc.*, *107*, 6069-6075 (1985).

134. L. J. Sannes and D. E. Hultquist, *Biochim. Biophys. Acta.*, *544*, 547-554 (1978).

135. D. E. Hultquist, L. J. Sannes, and D. A. Juckett, in *Curr. Topics Cellular Regulation*, *24*, 287-300 (1984).

136. H. Senn, F. Guerlesquin, F. Bruschi, and K. Wüthrich, *Biochim. Biophys. Acta.*, *748*, 194-204 (1983).

137. G. N. LaMar, D. L. Budd, and K. M. Smith, *Biochim. Biophys. Acta.*, *622*, 210-218 (1980).

138. I. Fita and M. G. Rossmann, *J. Mol. Biol.*, *185*, 21-37 (1985).

5

Metal-Porphyrin-Induced NMR Dipolar Shifts and Their Use in Conformational Analysis

Nigel J. Clayden, Geoffrey R. Moore,[*] and Glyn Williams[†]
The University of Oxford
Inorganic Chemistry Laboratory
South Parks Road
Oxford OX1 3QU United Kingdom

[*]Present affiliation: The School of Chemical Sciences, University of East Anglia, Norwich NR4 7TJ United Kingdom
[†]Present affiliation: Department of Chemistry, University College London, 20 Gordon Street, London WC1 0AJ, United Kingdom

1. INTRODUCTION

Metalloporphyrins are key components of a variety of biochemical
systems, including those responsible for O_2 transport and storage,
H_2O_2 utilization, energy transfer in the form of electron and proton
conduction, and the oxidation of various organic compounds. Great
progress has been made in defining chemical aspects of these processes and central to this progress has been the determination of
protein structures, usually by x-ray crystallographic methods, and
the construction of relatively simple porphyrin complexes to mimic
selected biochemical properties. In both of these areas NMR spectroscopy has a central role to play as a solution state method for
determining the molecular conformations of metalloporphyrins and
characterizing their electronic properties.

The purpose of this chapter is to present a theoretical
description of metalloporphyrin-induced NMR dipolar shifts and to
describe their use in conformational analysis. Other NMR paramagnetic methods, such as relaxation enhancement and contact shift
analysis, are considered fully elsewhere [1-10] and we shall only
refer to them when they have direct relevance to the analysis of
dipolar shifts.

2. THEORY

The NMR properties of a nucleus associated with a paramagnetic
metalloporphyrin are often dominated by the interaction between the
unpaired electron spin and the nuclear spin, which may lead to pro-
found changes in the chemical shift and relaxation times. In this
section we present a theoretical description of such paramagnetic
chemical shift perturbations from the viewpoint of their use in
conformational analysis. Metalloporphyrins may also give rise to
significant chemical shift perturbations by a diamagnetic mechanism,
namely, the ring current shift, and this must also be considered in
a conformational analysis. However, it can often be disregarded

because dipolar paramagnetic shifts can be measured directly. Nevertheless, we include a brief account of ring current shifts.

2.1. Paramagnetic Shifts

The electron-nuclear spin interaction can be separated into a through-space term, the dipolar or pseudocontact interaction, and a through-bond term, the Fermi contact interaction [11-13]. A general expression for the pseudocontact shift is given by Eq. (1) [12]:

$$\Delta pc = \frac{1}{3N}\left[\left\{\frac{1}{2}(\chi_{xx} + \chi_{yy}) - \chi_{zz}\right\}\frac{(3\cos^2\theta - 1)}{r^3}\right]$$
$$+ \frac{1}{2N}\left[\left\{\chi_{yy} - \chi_{xx}\right\}\frac{(\sin^2\theta\cos 2\phi)}{r^3}\right] \tag{1}$$

where χ_{ii} are the principal components of the magnetic susceptibility tensor and r, θ, and ϕ relate the nucleus of interest to the coordinate system of the susceptibility tensor (Fig. 1). In contrast the contact shift expression, Eq. (6) in [10], carries no explicit dependence on the spatial relationship of the metal ion to the nucleus of interest. Consequently, a conformational analysis necessitates a calculation of the pseudocontact shift and this requires only that the principal components of the magnetic susceptibility tensor and the orientation of this tensor with respect to the molecular frame be known.

Unfortunately, in most cases of interest these data are lacking. By restricting the description of the electronic state to a single thermally populated multiplet with an effective spin S, the susceptibility tensor components can be replaced by the EPR g values to give Eq. (2) [12]:

$$\Delta pc = \frac{\beta^2 S(S + 1)}{9kT}\left[\left\{\frac{1}{2}(g_{xx}^2 + g_{yy}^2) - g_{zz}^2\right\}\frac{(3\cos^2\theta - 1)}{r^3}\right.$$
$$\left. + \frac{3}{2}\left\{g_{yy}^2 - g_{xx}^2\right\}\frac{(\sin^2\theta\cos 2\phi)}{r^3}\right] \tag{2}$$

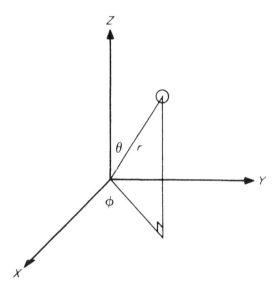

FIG. 1. The coordinate system of the pseudocontact shift equation
(1). The electron spin magnetic moment is taken to be localized at
the origin and the open circle represents the position of a nucleus
in the principal axis system. (Reproduced with permission from [14].)

This allows the use of EPR g values to calculate the pseudocontact
shift.

Although this widens the scope for pseudocontact shift calculations, it must be borne in mind that it is an approximation since
it neglects excited state and second-order Zeeman effects (see Sec.
4.4) [15,16]. Additional problems arise in the use of EPR g values
to represent the magnetic anisotropy. First, the g values themselves
are uncertain. Thus in frozen glasses at temperatures ≤100 K a wide
range of metal-porphyrin systems show a significant rhombic contribution, as evidenced by the observation of three distinct g values [7
and references therein], which is not present at room temperature
since the pyrrole βH protons are equivalent in the NMR spectra [6,7].
Second, the EPR g values can be exceedingly sensitive to the solution
conditions. Thus they may be solvent-dependent, as has been established for low-spin Co(II) porphyrins and is thought to be present in
axially ligated low-spin Fe(III) porphyrins [7 and references therein],
and affected by the porphyrin aggregation state [17].

The simplification of the pseudocontact expression to one
involving g-tensor values is particularly inappropriate for electronic states such as high-spin Fe(III) where a $^6A_1$ ground state is
encountered for here the g tensor is isotropic. However, the zero
field splitting (ZFS) of an electronic state with $S > 1/2$ can give
rise to a pseudocontact shift given by Eq. (3) [1,12]:

$$\Delta pc = \frac{28g^2\beta^2 D}{9k^2T^2}\frac{(3\cos^2\theta - 1)}{r^3} \tag{3}$$

where D is the ZFS parameter.

Equation (2) is also inappropriate for lanthanide ions because
at room temperature their g values are isotropic [18]. The susceptibility anisotropy that gives rise to pseudocontact shifts arises
when the symmetry of the ligand field is less than cubic, and
Bleaney [18] has presented a general equation for such lanthanide-
induced shifts:

$$\Delta pc = \frac{g^2\beta^2 J(J + 1)(2J - 1)(2J + 3)}{60(kT)^2 r^3}[D_z(3\cos^2\theta - 1) + (D_x - D_y)\sin^2\theta\cos 2\phi] \tag{4}$$

where D_x, D_y, and D_z are ligand field-splitting parameters.

This equation is a good approximation of the pseudocontact
shift of lanthanide ions, giving an accuracy of 10-20% at room temperature [19]. However, other terms do contribute to the pseudocontact shift, of which the most important is a T^{-3} term [19].
Also, in some cases low-lying excited states contribute to the Δpc.
These contributions are generally negligible at room temperature
but for Eu(III), for which $J = 0$, and Sm(III) they are the main
source of Δpc [18].

Equations (1), (2), and (4) may be simplified if the system
is axially symmetric, when the $(\sin^2\theta\cos 2\phi)$ term disappears (see
Sec. 4.1.4). This leads to a marked simplification of the procedure for determining conformations.

In the absence of the appropriate magnetic susceptibility
data, g values, ZFS, or ligand field-splitting parameters, pseudo-

contact shifts can still be calculated. In this empirical method a
nucleus with a known geometric relationship with respect to the metal
ion is used to parameterize the shift equation. Two assumptions are
implicit in this empirical method: first, that some part of the conformation is known and second, that the contact shift for the para-
meterizing nucleus is known. The additional assumption of axial
symmetry is also often made. Examples of this empirical approach
are given in Sec. 4.3.

The theoretical description of pseudocontact shifts given above
assumes that the unpaired spin resides on the metal ion. However,
this electron is delocalized to some extent into the porphyrin,
raising the question of whether ligand-centered pseudocontact shifts
(LCPS) need be considered [12]. Theoretical and experimental studies
[6,15,20-22] of the amount of unpaired spin located on carbon atoms
of the porphyrin range up to 2.0% of an unpaired electron for one
carbon atom. This will probably give rise to a substantial LCPS for
the ^{13}C nucleus in question but not for its attached proton owing to
the spherical nature of the s-orbital electron density. Resonances
of nuclei that do not form part of the porphyrin π-framework or
metal axial ligand will not be affected by LCPS.

2.2. Temperature Dependence of the Pseudocontact Shift

The most general form of the pseudocontact shift has no obvious temperature dependence [Eq. (1)] and it is the variation of the magnetic
susceptibility anisotropy with temperature that gives rise to the
temperature dependence of Δpc. Insofar as it is valid to approximate
the susceptibility expression to a g-tensor expression, we expect a
Curie law-type dependence of the pseudocontact shift, i.e., Δpc is
inversely proportional to the temperature [Eq. (2)]. By contrast,
for the case of an isotropic g tensor and significant ZFS we expect
the pseudocontact shift to be inversely proportional to the square
of the temperature [Eq. (3)]. Ln(III) pseudocontact shifts are also
expected to be approximately inversely dependent on the square of

the temperature [Eq. (4)]. A more complicated temperature dependence
may arise if these limiting cases are not valid. For example:

1. Significant SOZ due to the presence of low-lying excited
   states [16].

2. Spin state equilibria; high-spin/intermediate-spin/low-spin
   equilibria, as for ferricytochromes c' [23] and for
   $Fe(OEP)X_2ClO_4$ (X = substituted heterocyclic base), involving either a simple thermal mixture over noninteracting
   S = 5/2, 3/2, and 1/2 levels on one molecule or a quantum
   mechanical mixture caused by spin-orbit coupling [24].

3. Conformational changes in the molecule over the temperature interval studied which may either alter the electronic
   state of the metal ion or change the spatial relationship
   (r, θ, φ) of the nucleus to the metal center.

4. A reduction in the association constant of an intermolecular complex with increasing temperature.

In cases where the paramagnetic shift has not been separated
into its component pseudocontact and contact terms, the temperature
dependence will reflect changes in both terms. If the contact contribution is large, then we may observe a more complex temperature
dependence [1,7,16]. In this context it should be noted that ring
current shifts are not temperature-dependent [25].

2.3. The Porphyrin Ring Current Shift

Ring current shifts are through-space interactions, but although
they are usually represented by a dipolar field this is of questionable validity in view of the spatial extent of the charge distribution [26]. Nevertheless, various dipolar models have been
used for conformational analysis with intermolecular complexes and
proteins [27 and references therein]. The different dipolar models
represent the ring current by current loops placed above and below
the ring plane. These current loops are then replaced by dipoles
and it is here that the models differ [25,28] because the porphyrin

may be treated as a single dipole, five dipoles (placed at the four
pyrrole ring centers and the porphyrin center), or eight dipoles
(placed at the four pyrrole ring centers and the centers of the four
hexagonal rings of the macrocycle). The eight-dipole model of
Abraham et al. [28,29] appears to be the most satisfactory of these
models.

3.   STRUCTURAL AND MAGNETIC PROPERTIES OF METALLOPORPHYRINS

In this section we briefly describe certain structural and magnetic
properties of selected metalloporphyrins with the aim of establishing
which metal ions are generally best suited for conformational analysis.
Most metal ions have been inserted into porphyrins and in many cases
their NMR properties characterized. We are concerned with only a few
particular metals, specifically members of the first-row transition
metals, Zn(II) and the lanthanide series. These two series each provide a range of isomorphous porphyrins that includes diamagnetic,
pseudocontact shift, and paramagnetic relaxation reagents.

We only consider the porphyrin ring system although a number
of related systems are biochemically important, e.g., corrin and
chlorophyll. However, either the normal biochemical states of these
are diamagnetic or their paramagnetic states have not been well characterized. Recent reviews have appeared covering NMR of the naturally
occurring cobalt-substituted corrins [30] and magnesium-substituted
chlorophylls [31].

3.1. Stereochemistry of Metalloporphyrins

Metalloporphyrin stereochemistry has been extensively studied by
x-ray crystallography and the most significant influences shown to
be the complexing properties of the porphyrin and metal ion, and the
nature of the axial ligands [32-34 and references therein]. The

stereochemical parameters most affected by these factors are the
size of the porphinato core, as given by the radius of the central
hole, the Ct-N distance (Fig. 2), the effective size of the metal
ion, and the M-N bond lengths. In many porphyrins the M-N and Ct-N
distances are the same with the metal ion sitting in the center of
the hole but in some porphyrins M-N > Ct-N and the core is deformed
with the metal out of plane.

The extent of the out-of-plane deformation varies considerably
depending on the metal and its oxidation state, spin state, and
coordination number. The four-coordinate M^{n+}(TPP) complexes have
in-plane metal ions when M^{n+} is Fe(II), Co(II), Ni(II), Cu(II), and
Zn(II) but when it is Mn(II) it is out of plane. All the five-
coordinate metalloporphyrins have out-of-plane metal ions but in
six-coordinate systems M^{n+} is in plane or very nearly so. The most
substantial out-of-plane displacement is found with the six-coordinate

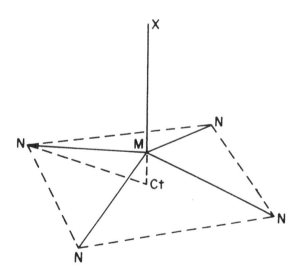

FIG. 2. The square-pyramidal coordination geometry for some metalloporphyrins. The metal is located at M and is bonded to the four
porphyrin nitrogens (N). An axial ligand bond is represented by the
solid vertical line. Ct is the center of the porphyrin ring. (Reproduced with permission from [33].)

Ln(III)(p-CH$_3$)TPP(diketonate) system where displacements of 1.6-
1.8 Å have been reported [35] (see Sec. 4.3).

The pattern of peripheral substituents around the porphyrin
does not significantly affect the structural parameters at the core
but the type of axial ligand may. The axial ligand effect is a consequence of changes in the metal ion radius upon change in spin state
and steric interactions between the axial ligand and porphyrin. Some
metalloporphyrins have been reported to have ruffled structures,
though whether this is an intrinsic structural feature or the result
of crystal-packing forces is not always clear. The heme of some low-
spin cytochromes is also deformed, having a saddle-shaped appearance
[36,37]. These cytochromes are c types with their heme covalently
bound to the protein via thioether groups formed by the condensation
of heme vinyls with Cys thiols, and it is these thioether groups
that may be partly responsible for the deformation.

Most of the stereochemical data listed above and described
fully in [32,34] were obtained by crystallographic methods. In solution various dynamic processes may occur to perturb these described
static structures. These dynamic processes have not generally been
well characterized but it is important to be aware of their possible
scope. For example, with metal ions displaced out of plane, porphyrin
inversion may occur relatively rapidly and this implies that the M-Ct
distance is not necessarily fixed. Other dynamic processes that may
occur are spin-state interconversions, axial ligand exchange, aggregation, and fluxional processes affecting the porphyrin macrocycle
and its substituents as well as the axial ligands. These are considered more fully in [7].

3.2. Electronic Properties of Metalloporphyrins

Conformational analysis using paramagnetic metalloporphyrins is
feasible only when the electronic properties of the metalloporphyrin
allow the observation of pseudocontact shifted resonances. Since
the electronic properties of the metalloporphyrin are determined

mainly by the metal ion, it is necessary to consider how the symmetry and spin state of the metal ion affect the electron spin relaxation times (T_{1e} and T_{2e}) and the magnitude and anisotropy of the magnetic susceptibility.

In most cases the electronic state of the metalloporphyrin can be described by a single spin state; however, exchange processes may produce spin state equilibria that are not generally capable of simple analysis. An example is the case of certain six-coordinate ferric hemoproteins which have pH-dependent spin states because of the ionization of their ligated H_2O [38]. In such cases the paramagnetic properties of the heme can only be usefully employed in conformational studies at pH values far removed from the pK. A related problem occurs with·certain five-coordinated ferricytochromes which have a quantum mechanically admixed S = 3/2, 5/2 state [23]. This is not a thermal equilibrium, so that it is generally not possible to operate under conditions such that one of the pure spin state types dominates.

The electron spin relaxation times are important because if they are too long, relaxation of the nuclei being observed is enhanced leading to considerable broadening of the resonances [see 2,4,5,7-9 and references therein]. In general, the condition for observing narrow resonances of a paramagnetic metalloporphyrin is $\tau_R \geqslant T_{1e}$ (where τ_R is the rotational correlation time of the molecule). Most small porphyrins and proteins have a τ_R of 10^{-9}-10^{-10} sec, e.g., the 12 kDal cytochrome c has a τ_R of 6 x 10^{-9} sec [39].

A value of T_{1e} for a given metalloporphyrin is difficult to predict a priori. However, a guide to expected values can be obtained from analyses summarized by Swift [2] and Horrocks [4]. These indicate that in general T_{1e}'s are sufficiently short for many of the first-row transition metals and the lanthanides to allow the observation of NMR resonances. The most important exceptions are those with A ground states where electron spin relaxation is inefficient; examples are high-spin octahedral d^5 [Mn(II) and Fe(III)] and f^7 [Gd(III)] ions, which are both powerful relaxation probes. It must be noted, however, that the expectation of long T_{1e} for metal ions with A symmetry is not always borne out experimentally because low-symmetry distortions, particularly ZFS, can provide efficient electron spin relaxation mechanisms.

The magnitude and anisotropy of the magnetic susceptibility determine whether large pseudocontact shifts will be generated by a given metalloporphyrin. A metal ion with A symmetry is expected to have a small g-tensor anisotropy and will consequently produce only small pseudocontact shifts according to Eq. (2); examples are high-spin Fe(III), Mn(III), and Ni(II). However, if a large ZFS is present, then substantial pseudocontact shifts may result from Eq. (3), as has been observed with certain high-spin Fe(III) systems [7,40]. For a degenerate ground state magnetic anisotropy may occur but the magnitude of this is often difficult to predict. In general, magnetic anisotropy is expected to be large for complexes with T ground states, a group that includes octahedral low-spin Mn(III) and Fe(III) and high-spin Fe(II).

The Ln(III) ions are extensively used in conformational analysis because, with the exception of the f^7 ion Gd(III), the paramagnetic Ln(III) ions have relatively short T_{1e}'s (a result of their strong spin-orbit coupling) and because they generate large pseudocontact shifts [41-43]. An indication of their strong pseudocontact-shifting ability is given by a comparison of Yb(III) and low-spin Fe(III) porphyrins (both are S = 1/2); the Yb(III) porphyrins have two to three times the pseudocontact-shifting ability of the Fe(III) porphyrins [35]. Thus, judged solely on the basis of electronic properties, Ln(III) porphyrins have considerable advantages over transition metal porphyrins. However, other considerations, such as the stereochemical differences and probable lack of biological function, counts against the Ln(III) porphyrins.

3.3. Summary

The properties described in the previous two sections are summarized in Table 1 for the metalloporphyrins best suited for conformational

TABLE 1

Summary of Structural and Magnetic Properties
of Selected Metalloporphyrins

Property	Details
Four-coordinate transition metals	
Stereochemistry	In-plane metal ions; good isomorphous series.
Diamagnetic	Ni(II) and Zn(II).
Pseudocontact shift	Co(II). [Fe(II) is expected to produce pseudocontact shifts but there is little experience with this.]
Dipolar relaxation	Co(II). [Cu(II) is less good [Ref. 17] and there is little experience of Fe(II).]
Five-coordinate transition metals	
Stereochemistry	Out-of-plane metal ions; probably not a good isomorphous series; may need to vary M^{n+} oxidation state to get a complete range of properties.
Diamagnetic	Co(III) [0.11], Zn(II) [0.33].
Pseudocontact shift	hs Fe(III) if large ZFS [0.45]; Co(II) [0.13].
Dipolar relaxation	hs Fe(III) [0.45]; hs Fe(II) [0.42].
Six-coordinate transition metals	
Stereochemistry	For low-spin systems the metal ions are generally in plane; good isomorphous system; may need to vary M^{n+} oxidation state to get a complete range of properties.
Diamagnetic	Co(III).
Pseudocontact shift	ls Fe(III), Co(II).
Dipolar relaxation	Cu(II), Co(II).
Tripositive lanthanide cations	
Stereochemistry	Out-of-plane metal ions; probably considerable deformation of the porphyrin core. X-ray structures required to determine if there is a good isomorphous series but expectations based on other complexes is that there will be.
Diamagnetic	La, Lu.
Pseudocontact shift	Pr, Nd, Sm, Eu give relatively small shifts with negligible broadening. Tb, Dy, Ho, Er, Yb give larger shifts with some broadening.
Dipolar relaxation	Gd.

Note: hs = high spin, ls = low spin. The figures in brackets are the M-Ct distances taken from the compilation of Scheidt [33].

analysis. If intermolecular complexes between simple metalloporphyrins and other molecules are to be studied, then the four-coordinate
transition metal series, or possibly the lanthanide series, are the
porphyrins of choice. However, many of the systems studied are hemoproteins where the choice of metalloporphyrin and attendant ligands
is not completely open, although it is not always restricted to iron
porphyrins either. Synthetic methods allow the replacement of the
entire metalloporphyrin, as in globins and b-type cytochromes [21,44,
45], or the central metal ion, as in c-type cytochromes [46]. Myoglobin has even been reconstituted with Yb(III) mesoporphyrin IX [47].
Nevertheless, despite the ability to change the metal, most NMR
studies of hemoproteins have retained the iron, though some have
used chemical methods to obtain more favorable properties. An
example of this is the addition of CN^- to high-spin ferrimyoglobin
which displaces the bound H_2O and causes the iron to go low spin
[15,20].

4. CONFORMATIONAL ANALYSIS

4.1. Practical Considerations

4.1.1. *Conformations in Solution*

The conformational analyses we describe have employed metalloporphyrin-
induced dipolar shifts to determine conformations of porphyrins, proteins, and interacting small molecule-porphyrin systems, all in solution. Therefore it is important to consider the dynamic nature of
molecular conformations in solution and to recognize the constraints
that this dynamism imposes on conformational analysis.

Rotation or oscillation about single bonds in a molecule, even
relatively rapid bond breakage and reformation, may occur readily,
and if such processes produce a number of different conformations
that are rapidly interconverting, the observed NMR parameters will
be averaged. Of course, these averaged parameters do not correspond
to a meaningful averaged conformation. It is this that complicates
conformational analysis. In principle, the problem can be overcome

because different parameters have different geometric dependencies. It is possible, therefore, to determine from the different averaged parameters the real molecular conformations and their populations provided sufficient independent parameters are determined. A fuller discussion of this topic is beyond the scope of this chapter and the reader is referred to [48]. However, we do stress the importance of confidence-testing derived conformations, either by the statistical procedures mentioned in Sec. 4.1.5, by the use of paramagnetic relaxation methods, or with coupling constant and NOE analyses.

An illustration of the marked conformational dynamism found in some proteins is the heme disorder of globins and cytochrome b_5 [44]. The protoheme group is asymmetric so that there are at least two modes of binding to the protein (Fig. 3). Both binding modes have been observed for hemoglobin, myoglobin, and cytochrome b_5, either

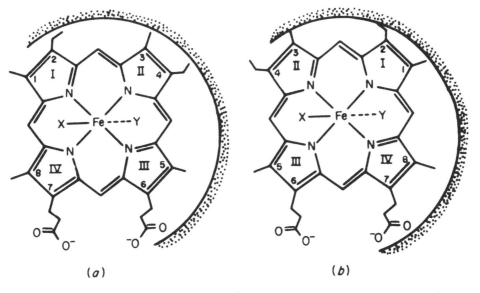

(a)   (b)

FIG. 3. Orientations of protoporphyrin IX in a hemoprotein. X and Y are the axial ligands. The orientations differ by a 180° rotation about the axis defined by the meso carbons linking pyrrole rings I and II, and pyrrole rings III and IV. (Reproduced with permission from [49], which is an adaption from [44].)

from heme reconstitution experiments or from experiments with freshly
prepared native protein. The two forms coexist in solution, usually
in slow exchange, and the form present in the solid state is usually
dominant in solution. However, in the case of cytochrome b_5, NMR
studies showed that the preferred orientation of the heme in solution
was opposite to the orientation suggested by x-ray studies [50],
although reanalysis of the x-ray data confirmed the NMR deduction
[51]. The heme orientation was determined by NMR from NOEs between
the heme and neighboring amino acids. Apart from illustrating one
facet of protein dynamics, these examples illustrate the importance
of assigning heme resonances and determining the heme orientation
independent from the x-ray structure, particularly if the heme resonances are to be used to help position the g tensor. Cytochromes c
have covalently bound heme and so do not exhibit the kind of disorder
shown in Figure 3.

4.1.2. *Separation of Diamagnetic and Paramagnetic Shifts*

The difference in chemical shift of a resonance between its value in
a metalloporphyrin complex or protein and its value in the porphyrin-
free molecular fragment or small, porphyrin-free peptides is its
conformation-dependent shift (CDS). This CDS is generally composed
of two terms, the induced diamagnetic shift and the induced paramagnetic shift, and in order to extract the paramagnetic contribution
from a measured CDS a correction for the diamagnetic term must be
made. This is done by selecting a suitable diamagnetic molecule that
has the same conformation as the paramagnetic molecule and taking the
chemical shift difference between corresponding resonances of these
two molecules as the paramagnetic shift. Table 1 lists appropriate
diamagnetic metalloporphyrins with their paramagnetic counterparts.

Generally this procedure works well but there are potential
problems. The main one is the question of how similar are the conformations of the diamagnetic and paramagnetic molecules? Coordination of axial ligands in hemoproteins influences the protein conformation, at least locally, and it is not certain that metal substitution can be made without some conformational perturbation, small

though that may be. Mitochondrial cytochrome c is a good example. There is a small redox state conformation change so that chemical shift differences, $\delta_{Fe(III)} - \delta_{Fe(II)}$, may not be solely paramagnetic shifts [36,52]. Also, comparison of the Zn(II)- and Co(III)-substituted cytochromes with the native Fe(II) cytochrome c reveals that there are small conformational differences that affect more than just the axial ligands and their immediate environments [53]. These diamagnetic shifts are generally $\leqslant 0.1$ ppm and so do not seriously affect the pseudocontact shift analysis given in Sec. 4.4.

Additional uncertainties are that some metals may have a sizable temperature-independent paramagnetism [such as some Co(III) complexes], or cause different electric field effects as a result of different charges, or vary the magnitude of the heme ring current. None of these possibilities have been fully assessed although they do not appear to be important for the metal-substituted cytochromes c [53]. What may be important in some studies is that resonances of the metalloporphyrin itself are very sensitive to metal substitution, even when the metalloporphyrin is diamagnetic, and sizable shifts may occur [31].

Contrasting the procedure for obtaining experimental paramagnetic shifts with the analogous approach for obtaining experimental ring current shifts is instructive. In this case the CDS for a resonance in a diamagnetic metalloporphyrin complex or protein is usually taken to be the shift caused by the porphyrin ring current (and any other ring current centers in the molecule). This assumes that other diamagnetic shift effects are negligible, an assumption that is often not valid [54,55], and this is one of the reasons why conformational analyses using metalloporphyrin-induced diamagnetic shifts are usually less satisfactory than the analogous paramagnetic analyses.

4.1.3. *Separation of Pseudocontact and Contact Shifts*

The important step of separating an experimental paramagnetic shift into its component pseudocontact and contact shifts is generally straightforward. In most cases the observed resonances are not from ligands of the metal ion and therefore do not experience a contact shift. If the resonances come from a ligand to the metal, and therefore might be expected to experience a contact shift, the separation can only be done by calculating one of the expected shifts. With lanthanide ions, theoretical procedures for separating the shifts have been proposed and in some cases they work satisfactorily [41,42]. However, their applicability to Ln(III) porphyrins has not been tested.

4.1.4. The Favorable Case of Axial Symmetry

Pseudocontact shift analysis is considerably simplified under conditions of axial symmetry. This occurs when $\chi_{xx} = \chi_{yy} \neq \chi_{zz}$. Taking $g_{zz} = g_\parallel$ and $g_{xx} = g_{yy} = g_\perp$, Eq. (2) becomes:

$$\Delta_{pc} = \frac{\beta^2 S(S + 1)}{9KT} (g_\parallel^2 - g_\perp^2) \frac{(3\cos^2\theta - 1)}{r^3} \qquad (5)$$

Equation 4 is similarly simplified.

These simplified equations can be used as they are, provided the appropriate magnetic susceptibility or g-tensor data are available, but their form allows a ratio method to be used that eliminates the need for magnetic susceptibility data. Thus the ratio of the pseudocontact shift for two nuclei in the same molecule or complex is

$$\frac{\Delta_{pc}^i}{\Delta_{pc}^j} = \frac{(3\cos^2\theta_i - 1)/r_i^3}{(3\cos^2\theta_j - 1)/r_j^3} \qquad (6)$$

Provided sufficient experimental shift ratios are available, possible conformations can be readily distinguished.

The validity of this procedure is not always simple to establish. With the lanthanide series its reliability can usually be determined by the consistency of experimental shift ratios for different lanthanides [41,42]. This comparative approach is not possible with transition metal porphyrins and so either EPR data or confidence tests (Sec. 4.1.5) must be used. The latter procedure assumes axial symmetry to derive a conformation and then demonstrates that the conformation is consistent with all other available data—a procedure that is not always satisfactory.

The origin of axial symmetry in nonporphyrin lanthanide complexes has been widely studied because it is not expected from x-ray structures that axial symmetry should be common [41,42 and references therein]. The explanations put forward for the observed effective axial symmetry are based on considerations of the dynamic properties of the systems being studied: effective axial symmetry results from rapid exchange between different lanthanide-ligand complexes and from dynamic processes within a given complex. Similarly, the effective axial symmetry of some transition metal porphyrins at room temperature has been ascribed to dynamic processes that average the in-plane anisotropy observed at low temperature [6,7]. However, there is no such averaging for hemoproteins and thus for most proteins the full rhombic Eq. (2) should be used.

4.1.5. Confidence Testing-Derived Conformations

The best method of assessing the accuracy of a conformation proposed from a pseudocontact shift analysis is to analyze other NMR parameters with different geometric dependencies. Paramagnetic relaxation studies with the appropriate metalloprotein (Table 1) are often done and, with proteins at least, interpretation of NOE and spin-spin coupling constants may be useful. Analysis of the metalloporphyrin-induced ring current shifts is occasionally done but since these shifts usually have the same geometric dependence as the pseudocontact shifts, they are not generally of much use in assessing the accuracy of a pseudocontact shift-derived conformation.

Another approach is to use the reliability factor (R):

$$R = \frac{\sum_i (\Delta^i_{obs} - \Delta^i_{calc})^{1/2}}{\sum \Delta^i_{obs}} \qquad (7)$$

where Δ^i_{obs} and Δ^i_{calc} are the observed and calculated shifts, respectively, and the sum is extended over all observed resonances [56,57].

Δ^i_{calc} may be obtained from a model built on the basis of the pseudo-contact shift analysis or from a completely independent model, such as an x-ray structure. The confidence with which one set of calculated values is preferred to any other set (i.e., which set agrees best with the observed values) can be determined from the ratio of their R values in conjunction with standard statistical tables [57, 58]. However, it should be noted that such an approach is only useful if the observed shifts are large or at least of similar magnitude to the experimental errors. When the observed shifts are small, the reliability factor may be dominated by the contributions of a few shifts.

4.2. Intermolecular Complexes

Biological interest in intermolecular complexes involving metallo-porphyrins centers on two main topics: complexes involving porphyrins themselves and complexes involving heme enzymes. We consider each in turn.

Complexes involving porphyrins are of interest partly because they may be models for oxidative heme enzymes but also because they may be important in the biosynthetic, degradative, and transport stages of porphyrin utilization. Conformational studies have been made of the complexes formed with 2,4,7-trinitrofluorenone [59], purines [17], steroids [60], and quinones [61]. In their porphyrin studies, Williams and coworkers [17,59,60] used the procedure they advocate for lanthanide complexes [41,62], namely, the combined use of a diamagnetic probe [such as Ni(II)MPDE], a relaxation probe [such as Cu(II) or Fe(III)MPDE], and a shift probe [Co(II)MPDE; Co(II)MPDE also enhances relaxation]. By taking the difference between the shifts induced by Ni(II) and Co(II) porphyrins as the pseudocontact shift, and using the ratio method to eliminate unknown constants (Sec. 4.1.4), a computer search procedure could then be used to generate families of structures. A unique structure was not found in any case but many of the structures could be eliminated by

taking account of the Co(II) and Fe(III) relaxation data. Even so,
a number of possibilities remained.

A typical structure of the Co(II)MPDE-caffeine complex in
$CDCl_3$ is shown in Figure 4. All of the structures consistent with
the experimental data were of the same general plane-to-plane type
as that in Figure 4. The major uncertainty with such structures is
the intermolecular separation. For Co(II)MPDE-caffeine, solutions
were found with the distance between the metal and the C5 atom of
caffeine varying from 3.8 to 4.4 Å. This spread may reflect the
coexistence of a range of complexes with varying intermolecular separations or it may indicate that insufficient parameters were available to precisely define the separation. Including the Ni(II)MPDE,
ring current shifts in the computer search may aid the procedure but
are unlikely to significantly improve the determination of intermolecular separations, as the porphyrin-quinone complexes described
below demonstrate. The main assumptions in the analyses of Co(II)MPDE
shifts described by Williams et al. [17] is that contact contributions
are negligible and the complexes can be treated as having axial symmetry. The fact that chemically reasonable structures could be obtained supports these assumptions but direct support was also obtained

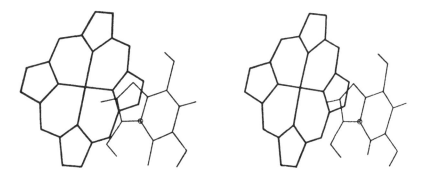

FIG. 4. Stereoscopic view of a possible solution structure for the
caffeine-Co(II)MPDE complex in $CDCl_3$. The C5 atom of caffeine is
represented by a circle. (Reproduced with permission from [17].)

using EPR. In the case of caffeine, coordination of the N9 atom to the Co(II) (the most likely path for a contact shift mechanism) could be ruled out, and in all the complexes of Co(II)MPDE examined $g_\perp \geqslant g_\parallel$ with little evidence of rhombic distortion.

Moreau et al. [61] studied the complexes of hydroxyferriprotoporphyrin IX (HFPP) and uroporphyrin I (Uro) with quinine and chloroquin because HFPP produced by the degradation of hemoglobin may be the site of action for the quinine antimalarial drugs. HFPP polymerizes in aqueous solution, and although its interaction with quinine causes resonances of the latter to be shifted, aggregation effects and lack of a suitable diamagnetic control make quantitative analysis difficult. Therefore, Moreau et al. [61] chose to determine the structure of the quinine-uroporphyrin complex as a model for the HFPP complex. Uro has the advantage that it does not aggregate in aqueous solution until relatively high concentration. Also, since the HFPP studies indicated that the metal was not important for complex formation, metal-free uro was used so that the ring current shifts could be analyzed. Using the ring current model developed by Abraham et al. [27,28] and taking ratios of chemical shifts, a plane-to-plane structure was proposed. This is similar to the structures determined for other systems, such as Figure 4, and like these earlier studies the intermolecular separation is not well defined since satisfactory fits to the data were obtained for separations of 2.5-4.5 Å.

Heme enzymes, such as cytochrome P-450 and horseradish peroxidase (HRP), oxidize a wide range of organic compounds and a key step in their reaction cycle is formation of an enzyme-substrate complex. It is this step that should be particularly amenable to study using porphyrin-induced dipolar shifts although, surprisingly, only one such study has been reported.

Binding of the substrate indolepropionic acid (IPA) to HRP has been studied by the procedures described previously for caffeine-porphyrin complexes [63]. Like those complexes the interaction is not particularly strong and so the system is in fast exchange. The

diamagnetic CO-bound ferrous form of HRP was taken as the diamagnetic
control and the resonance shifts of IPA bound to the high-spin ferric
form of HRP used to obtain the pseudocontact shifts. Analysis of the
pseudocontact shifts and relaxation effects define a structure in
which the indole ring is located 9-10 Å from the heme iron (Fig. 5).
This structure is very different from those of the intermolecular
complexes and it seems probable, at least for HRP, that plane-to-
plane complexes are unlikely to be formed.

4.3. Small Porphyrins

Although paramagnetic metalloporphyrins have been extensively studied
by NMR spectroscopy, there have been few reports of the dipolar shift
being used in conformational analysis. In contrast to this, there
have been numerous papers dealing with the electronic structure as
deduced from the spin delocalization implied by contact shifts [3,4,7
and references therein], and this first requires that the dipolar
shifts be calculated. Consequently, it is implicit in the spin delocalization analyses that the conformation is defined and the magnetic
anisotropy known. The first assumption is generally true since
usually the porphyrins are rather simple with known single-crystal
structures. Magnetic anisotropy data in the form of the principal
components of the susceptibility tensor are rarely available but in
general g-tensor data can be substituted. In the absence of susceptibility or g-tensor data the dipolar shift may be calculated empirically by assuming a negligible contact shift on a resonance, e.g.,
a mesophenyl meta proton.

Dipolar shifts have been used to demonstrate axial ligation in
a σ-alkyl iron porphyrin [64]. The magnetic anisotropy of the TPP
Fe(III)-nC$_4$H$_9$ complex was represented using the EPR g values found
at 77 K in a frozen solution. Geometric factors were calculated for
the γCH$_2$ and δCH$_3$ of the σ-n butyl group assuming an Fe-C bond length
of 1.8 Å. Reasonable agreement was then found between the calculated
dipolar shifts and observed paramagnetic shifts, although it should

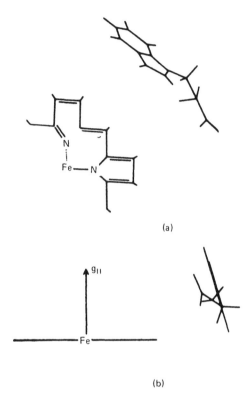

(a)

(b)

FIG. 5. Structure of the indolepropionic acid-horseradish peroxidase
complex: (a) projection onto the heme plane and (b) projection onto
the plane containing the axis and perpendicular to the indole ring.
The outline of the porphyrin ring in the XY plane is in an arbitrary
orientation. The shift ratios and distance ratios (measured from
relaxation effects) are:

Proton	Shift ratios		Distance ratios	
	Experimental	Computed	Experimental	Computed
H4	100	100	100	100
H2	21	21	101	103
H7	29	29	103	104
H5	~ 70	88	-	104
H6	~ 35	66	-	107
CH_2	79	{ 113 63	108	{ 95 112
CH_2CO_2H	74	{ 105 54	112	{ 90 108

The maximum fully bound pseudocontact shift for H4 was -0.72 ppm.
(Reproduced with permission from [63].)

be noted that no allowance was made for contact shifts because it was assumed that they had been attenuated by the intervening bonds. It was concluded from this agreement that the alkylated complexes possess a σ-carbon-iron bond.

Axial ligation for the complexes Fe(III) TPPX [X = $SO_3CF_3^-$, ClO_4^-, and $C(CN)_3^-$] has been examined in the light of dipolar shift calculations [65]. The main problem is uncertainty regarding the magnetic anisotropy of the admixed spin state of Fe(III)TPPX (S = 5/2, 3/2). This was overcome by estimating a dipolar shift for the orthophenyl proton in Fe(TPP)ClO_4 (which has substantial S = 3/2 character) and assuming this value provides an approximate value for the dipolar shift for the orthophenyl proton in the $SO_3CF_3^-$ and $C(CN)_3^-$ adducts (which have a lesser S = 3/2 character). This is valid providing the ligands do not induce a significant ZFS. Maximum dipolar shifts were then calculated for an ion-paired ligand by placing the ligand nuclei of interest on or near the Z axis. In all three cases the maximum dipolar shift calculated was significantly less than the observed paramagnetic shift and therefore the ligand was assumed to be coordinated and contact shifts invoked to explain the difference.

The scope and applicability of conformational analysis in small porphyrins increases as the complexity of the porphyrin increases. A clear example is provided by the "capped" porphyrins (Fig. 6) where the aliphatic chains joining the cap to the porphyrin allow sufficient constraint to prevent large-scale movements, such as would be present in "picket fence" and "basket handle" porphyrins [66], but still allow enough flexibility to alter the separation between the cap and the porphyrin. Since the ability of the iron(II) "capped" porphyrins (CP) to bind oxygen depends on the cavity between the porphyrin plane and capping benzenoid ring, a knowledge of the solution conformation is invaluable. The cobalt(II) capped porphyrins are preferable to study rather than the iron(II) porphyrins because these are low-spin Co(II) with the unpaired electron in the d_Z^2 orbital, which minimizes the contact contribution to the paramagnetic shifts. A good diamagnetic reference is the zinc-substituted CP [67].

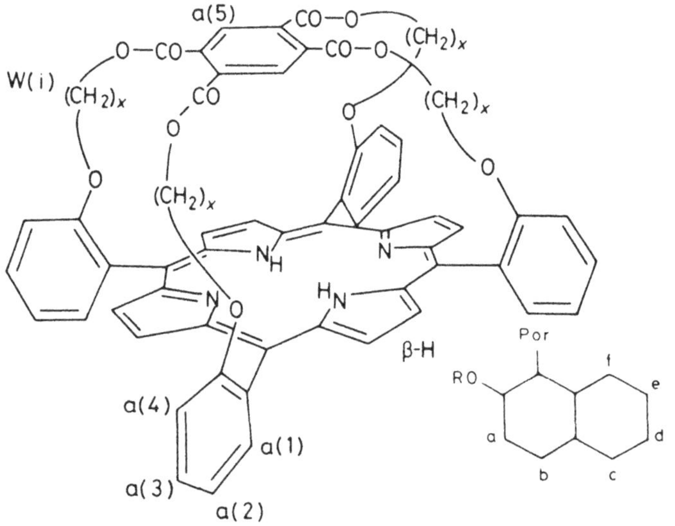

FIG. 6. Capped porphyrins: CP2, x = 2; CP3, x = 3; CP2(N), x = 2 and mesophenyl groups replaced by naphthyl groups. The observed (O) and computed (C) pseudocontact shift ratios for the aryl protons are:

	Co(II)CP2			Co(II)CP3			Co(II)CP2(N)	
Proton	O	C	Proton	O	C	Proton	O	C
a1	100	100	a1	100	100	a	111	113
a2	43	46	a2	44	46	b	100	100
a3	42	41	a3	49	41	c	39	71
a4	51	46	a4	66	46	d	6	20
						e	-100	-33
						f	137	121

The large discrepancies for some of the Co(II)CP2(N) protons is probably because there are deviations from the idealized values used for the naphthyl-porphyrin geometry. The a5 pseudocontact shifts at 27°C were CoCP2, 19.76 ppm; CoCP3, 21.90 ppm; CoCP2(N), 19.40 ppm. (Reproduced with permission from [68].)

Dipolar shifts were calculated empirically by assuming the mesophenyl para proton shift was purely dipolar and axial in character [68]. This gave magnetic anisotropies $(g_{\parallel}^2 - g_{\perp}^2)$ of -9.3 for CoCP2, -9.8 for CoCP2(N), and -8.5 for CoCP3, in contrast to -8.2 for Co[O-CH$_3$TPP]. Following McGarvey [69], these differences were related to low-

symmetry perturbations to the C_{4V} symmetry, these being more pronounced for the rigid CoCP2(N) than for the more flexible CoCP3.
Close agreement was found between the calculated dipolar shifts and
the observed paramagnetic shifts for the other mesoaryl protons
(Fig. 6), which have a known geometric relationship to the metal
center. Having demonstrated the validity of the magnetic anisotropies noted above, the cap-porphyrin separation was then obtained
from the cap proton (a5) paramagnetic shift. These were found to
be CoCP2 3.65 Å, CoCP2(N) 3.8 Å, and CoCP3 3.4 Å. The greater
aliphatic chain length in the CP3 series causes a decrease in the
cavity size because the additional methylene groups fold into the
cavity and pull the capping benzene and porphyrin together. Support
for this conclusion regarding the cavity size was provided by the
resonance linewidths. Taking the cobalt-pyrrole βH distance as
5.3 Å, the cap-porphyrin separation was found to be 4.3 Å for CoCP2
and 3.95 Å for CoCP3. Although the separation is greater than that
indicated by the dipolar shifts, the relative order is maintained.

The extent of the out-of-plane displacement of Ln(III) ions
complexed with porphyrins at -21°C has been determined by pseudocontact shift analysis [35] (Fig. 7). This analysis assumed axial
symmetry with the principal axis normal to the porphyrin plane and
negligible contact shift on the aryl resonances. Also it used
structural parameters for the TPP from x-ray structures where the
aryl ring planes are normal to the porphyrin plane. Although this
is a crude model, since it ignores phenyl ring oscillations and
major deformations of the porphyrin, there is good agreement between
the observed and calculated shift ratios for the aryl protons (Fig.
7). The poor agreement with the pyrrole resonances (d) is consistent
with the expectation of sizable contact shifts for these resonances.
The aryl proton shift ratios indicate that the lanthanides of the
Eu(III) and Yb(III) complexes are displaced from the porphyrin
plane by 1.8 and 1.6 Å, respectively, which are equivalent to Eu-N
and Yb-N distances of 2.69 and 2.56 Å, respectively.

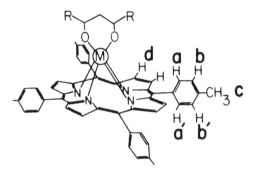

FIG. 7. Schematic representation of Ln(III)(p-CH$_3$)TPP(diketonate).
^{1}H NMR spectra of the Eu(III) and Yb(III) complexes in CDCl$_3$ at -21°C
gave the following experimental shift ratios. These are compared to
the computed ratios for complexes with varying M-Ct distances.

Proton	Experimental shift ratios		Computed shift ratios for M-Ct (Å)	
	Eu(III)	Yb(III)	1.8	1.6
Ha	100	100	100	100
Hb	31.4	31.9	31.2	31.7
CH$_3$	12.8	13.6	12.9	13.7
Ha^0'	-6.1	2.5	-4.9	-1.4
Hb^0'	6.8	9.2	8.3	10.0
Pyrrole(d)	-18.9	69.5	55.5	64.1

The experimental shifts for the Ha resonances of the Eu(III) and
Yb(III) complexes were -4.9 and -24.1 ppm, respectively. (Reproduced with permission from [35].)

4.4. Hemoproteins

Although many different kinds of hemoprotein exist, high-resolution
NMR studies have been confined to two classes: the O$_2$ binding globins
and the electron carrying cytochromes. The former proteins are high
spin in their native ferric and unliganded ferrous states whereas the
better characterized of the latter proteins are low spin. Inevitably,
this difference in spin states has led to a different emphasis being
placed on their study by NMR, with pseudocontact shift analysis being
a central feature of many cytochrome studies but little such analysis

being reported for native globins. Dipolar shift analyses for the
low-spin complexes metmyoglobin cyanide [Fe(III)] and oxy- and
carbonmonoxymyoglobin [Fe(II)] have been reported [15,20,70], but
before they can be properly assessed independent resonance assignments are required. In view of this we restrict our further discussion to cytochromes.

The first detailed analysis of the pseudocontact shifts experienced by resonances from amino acids located near the heme of a
protein was reported by Keller and Wüthrich [71] for ferricytochrome
b_5. The structure of this monoheme protein of 93 residues had been
determined by x-ray crystallography [72] and the principal values of
the g tensor measured in a frozen solution with EPR [73]. Keller
and Wüthrich attempted to use these data to interpret the NMR spectra
but because of the lack of single-crystal EPR data the orientation
of the g-tensor axes relative to the molecular axes was undefined.
To overcome this problem an orientation of the g tensor was found
which was reasonably consistent with the number of methyl resonances
shifted to high field ($\delta \leqslant 0$ ppm) and their temperature dependencies.
However, this analysis relied on calculations using an unrefined
coordinate set and assumed that the pseudocontact shifts were inversely dependent on temperature. In the light of the cytochrome c
data discussed below, it is likely that these procedures introduce
significant errors and thus it is not surprising to find that these
early calculations are not consistent with some of the recent measured pseudocontact shifts [50]. One of the original aims of the
cytochrome b_5 analysis was to determine the polypeptide conformation
around the heme and by ring current shift calculations with the
modified coordinates of cytochrome b_5, taking account of both the
heme and amino acid ring currents, this has been shown to be not
significantly different in the crystal and solution states. This
is in agreement with NOE measurements [50].

A similar approach to the above was pursued in the case of the
monoheme protein tuna cytochrome c, which consists of 103 residues.
In this case, single-crystal EPR data [74,75] and two independent

sets of x-ray coordinates [35] were available, as well as many
resonance assignments in both the paramagnetic ferric and diamagnetic ferrous states that are independent of the three-dimensional
structure [76,77]. The procedure used was as follows [14]:

1. A subset of 21 resonances was selected to optimize the
    g-tensor orientation. None of these belonged to ligands
    of the iron and so did not experience contact shifts. In
    addition, their redox state shifts (i.e., chemical shift
    difference between ferricytochrome c and ferrocytochrome c,
    Δ_{pc}^{obs}) were not greatly influenced by conformation changes.

2. The measured g values were corrected for the first-order
    Zeeman (FOZ) and second-order Zeeman (SOZ) effects (Table
    2).

3. The optimal g-tensor orientation was determined for the
    subset of 21 resonances by comparison of the least squares
    differences between the calculated and observed shifts
    [i.e., by minimizing $\Sigma(\Delta_{pc}^{cal} - \Delta_{pc}^{obs})^2$] for various orienta-
    tions.

The main conclusions from this analysis are illustrated in
Tables 3 and 4. They are:

1. FOZ and SOZ corrections to the g values improve the agree-
    ment between observed and calculated shifts but not by as
    much as reorienting the g tensor.

TABLE 2

Principal g-Tensor Components of Ferricytochrome c

	g_x	g_y	g_z
Measured at 4.2 K [74,75]	1.25	2.25	3.06
FOZ corrected [16]	1.28	2.22	2.94
FOZ + SOZ corrected [14,16]	1.95	2.59	3.24

Source: Reproduced with permission from [14].

TABLE 3

Optimization of g-Tensor Orientation

	Unrotated			Rotated		
	Low	FOZ	+SOZ	Low	FOZ	+SOZ
Inner	8.4	7.5	6.8	3.7	2.2	1.4
Outer	6.5	5.4	4.4	3.4	1.9	1.1

Note: Low, FOZ, and +SOZ refer to the three sets of g values given in Table 2 and are the values measured at 4.2 K, FOZ corrected, and FOZ + SOZ corrected values, respectively. Inner and outer refer to the two independent x-ray structures of tuna ferricytochrome c [36], and the figures in the table are the values of $\Sigma(\Delta_{pc}^{calc} - \Delta_{pc}^{obs})^2$ for the subset of 21 resonances given in Table 4. The $\Sigma(\Delta_{pc}^{calc} - \Delta_{pc}^{obs})^2$ values for the outer structure, FOZ + SOZ corrected g values, and the full set of 95 resonances given in [14] are 20.6 and 4.5 for the unrotated and rotated g tensors, respectively.
Source: Adapted from [14].

2. The EPR determined g-tensor orientation obtained at 4.2 K does not give the best fit to the observed pseudocontact shifts. The best-fit g-tensor orientation has g_z 11° from the heme normal (Fig. 8). For only one resonance of the subset does the low-temperature g-tensor orientation give better agreement than the reoriented g tensor.

3. The two crystal structures give significantly different results.

The main conclusion to be drawn about the conformation of cytochrome c is that the NMR data are generally consistent with the x-ray structure except for three regions of the protein. These differences, which are primarily due to intermolecular crystal-packing constraints, are discussed elsewhere [80]. NOE and coupling constant measurements support the structural model based on the pseudocontact shifts [81] and, as expected, so do ring current

TABLE 4

Comparison of Redox State Shifts and Calculated
Pseudocontact Shifts for Tuna Cytochrome c

Proton(s)	Measured shift (ppm)	Calculated shifts (ppm)	
		Rotated g tensor	Unrotated g tensor
Ala 4β	-0.04	-0.10	-0.13
Ala 15β	0.77	0.70	0.25
Thr 19γ	1.11	1.16	1.16
Val 20γ$_1$	0.95	0.78	0.50
Val 20γ$_2$	0.61	0.56	0.40
Leu 32δ$_1$	2.55	2.75	2.05
Leu 32δ$_2$	2.16	2.05	3.08
Leu 35γ$_1$	-0.21	-0.18	-0.06
Leu 35γ$_2$	-0.58	-0.44	-0.39
Ala 43β	-0.08	-0.09	0.20
Leu 64δ$_1$	-0.78	-0.59	-0.33
Leu 64δ$_2$	-1.25	-0.86	-0.70
Met 65ε	-0.20	-0.21	-0.14
Leu 68δ$_1$	-1.72	-1.68	-1.39
Leu 68δ$_2$	-3.11	-3.34	-2.18
Ala 83β	0.18	0.24	0.49
Leu 94δ$_1$	-1.00	-1.23	-1.44
Leu 94δ$_2$	-1.66	-2.24	-2.19
Ala 96β	-0.20	-0.25	-0.26
Leu 98δ$_1$	-0.30	-0.63	-0.98
Leu 98δ$_2$	-1.23	-1.70	-1.80

Note: The measured shifts are the chemical shift differences,
δFe(III) - δFe(II), and the calculated shifts were obtained using
the outer coordinate set of ferricytochrome c [36] with Eq. (2)
and either the EPR-determined g-tensor orientation (unrotated set)
or the NMR-determined orientation (rotated set) with the FOZ and
SOZ corrected g values (Table 2) [14].
Source: Adapted from [14].

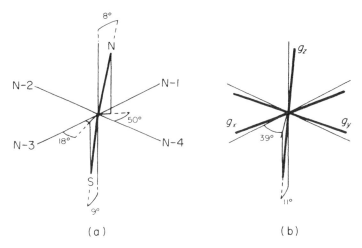

FIG. 8. Comparison of (a) the geometry of the heme ligands of ferricytochrome c (taken from the outer coordinate set of [36]), with (b)
the g-tensor orientation determined by fitting the pseudocontact
shifts calculated with the FOZ- and SOZ-corrected g values of Table 2
to the shifts observed at 27°C. The g-tensor orientation is significantly different from that measured by EPR at 4.2 K [74]. g_z is
displaced from the pyrrole N-plane normal approximately in the direction of the Met 80 Sδ-Fe bond vector. This is consistent with the
proposal that the g-tensor anisotropy of hemoproteins is partly a
reflection of axial ligand-porphyrin interactions [6,7,15,20,78,79].
(Reproduced with permission from [14].)

calculations for ferrocytochrome c [N. J. Clayden, G. R. Moore, S. J.
Perkins, and G. Williams, unpublished data].

The observation that the magnitudes of the g-tensor components
are not critical in determining the pseudocontact shifts is borne out
by recent studies of cyanoferricytochrome c in which the Met 80 ligand
of the iron is replaced by CN⁻ [Z. X. Huang, G. R. Moore, G. Williams,
and R. J. P. Williams, unpublished data]. This replacement does not
greatly perturb the protein structure although it does perturb the
heme electronic structure. The principal g values measured in frozen
solution at 10 K are 0.93, 1.89, and 3.45 [82]. Yet despite the variation in g values, many of the resonances from heme-packing residues
are little affected, e.g., Δ_{pc}^{obs} for Ala 15β and Thr 19γ are 0.65 and
1.14 ppm, respectively (compare with Table 4).

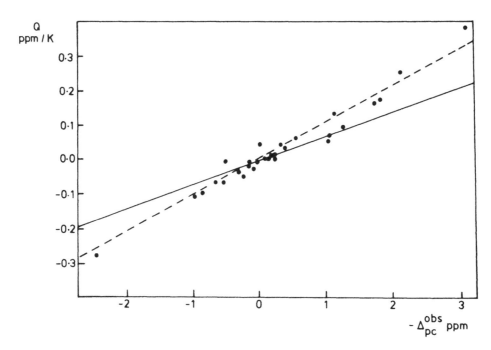

FIG. 9. Plot of the temperature coefficient of the chemical shift Q
(change in chemical shift between 27 and 50°C) of 30 amino acid
methyl resonances of horse ferricytochrome c at pH 5.5 against their
observed pseudocontact shift at 27°C. The solid line is the theoretical line drawn with a gradient of [1 - (300/323)] through the point
$Q = 0.0$, $\Delta_{pc}^{obs} = 0.0$ and it assumes Δpc is inversely dependent on the
temperature. The broken line, which is a better fit to the data,
was fitted by eye.

A complete analysis of pseudocontact shifts has to account for
their temperature dependence. Although Eq. (2) indicates that there
should be a simple inverse temperature dependence, this is rarely
observed [6,52]. Figure 9 emphasizes this point; the plot illustrates that there is a correlation between the Δ_{pc}^{obs} and its temperature dependence, but the Δ_{pc}^{obs} decreases more rapidly with increasing
temperature than predicted by Eq. (2). There are many reasons why
this may be so (Sec. 2.2).

Figure 10 illustrates the qualitative use of Δ_{pc}^{obs} for conformational analysis. Even though many of the residues of *Candida* ferri-

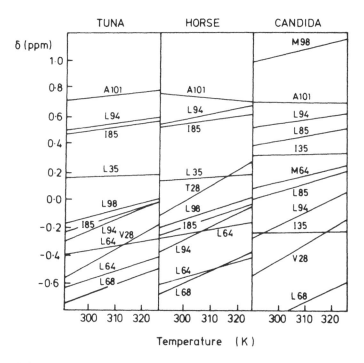

FIG. 10. The temperature dependence of the chemical shifts of selected methyl resonances of tuna, horse, and *Candida krusei* ferricytochromes c at pH 5.5. These temperature-dependent shifts are almost entirely caused by the temperature dependence of their pseudocontact shift contributions because resonances of the diamagnetic ferrocytochrome c and cobalticytochrome c are not markedly temperature-dependent [52,53].

cytochrome c are different from those of horse and tuna cytochromes c, corresponding methyl groups have similar Δ_{pc}^{obs} and temperature dependencies (compare Met 98 with Leu 98, Val 28 with Thr 28, Met 64 with Leu 64, and Ile 35 with Leu 35), indicating that their positions with respect to the heme are similar.

5. CONCLUSIONS AND OUTLOOK

Metalloporphyrin-induced pseudocontact shifts have a firm theoretical base that enables them to be used for conformational analysis. The major problem in this is defining the principal values and directions

of the magnetic susceptibility tensor, although there are empirical
methods for overcoming this problem. These methods generally rely
on the system being axially symmetric, which, although a property
of many small porphyrin complexes, is not common among hemoproteins,
which are largely rhombic. This, coupled with the need for detailed
spectral assignment, probably explains why there is a dearth of hemoprotein studies. However, this situation will change since with the
continuing development of NMR methodology characterization of complex
NMR spectra is becoming relatively straightforward.

There are two general types of NMR-based conformational study:
comparative studies with preexisting structural models, usually x-ray
crystal structures, and a priori structure determinations. The
pseudocontact shift method is a powerful procedure for the former
type of study but has a more limited application for the latter type.
This is a common limitation to all solution state techniques but one
that may be overcome, at least in principle, by the combined use of
different methods, such as NOE and coupling constant analysis as well
as paramagnetic relaxation effects. Thus, our expectation is that
metalloporphyrin pseudocontact shift analysis will be used increasingly to tackle problems of protein structure-function relationships,
particularly for substrate binding to enzymes.

ACKNOWLEDGMENTS

We are indebted to Prof. R. J. P. Williams for his advice on all
topics covered in this chapter. G. R. Moore also thanks the Science
and Engineering Research Council for an Advanced Fellowship, and
N. J. Clayden thanks Shell, Thornton Research Centre for financial
assistance.

ABBREVIATIONS

CDS	conformation-dependent shift
CP	capped porphyrin
EPR	electron paramagnetic resonance
FOZ	first-order Zeeman effect
HFPP	hydroxyferriprotoporphyrin IX
HRP	horseradish peroxidase
IPA	indole-propionic acid
LCPS	ligand-centered pseudocontact shift
MPDE	mesoporphyrin IX dimethyl ester
NMR	nuclear magnetic resonance
NOE	nuclear Overhauser effect
OEP	octaethylporphyrin
SOZ	second-order Zeeman effect
TPP	tetraphenylporphyrin
Uro	uroporphyrin
ZFS	zero field splitting

REFERENCES

1. J. P. Hesson, in *NMR of Paramagnetic Molecules* (G. N. LaMar, W. deW. Horrocks, Jr., and R. H. Holm, eds.), Academic Press, New York, 1973, p. 2.

2. T. J. Swift, in *NMR of Paramagnetic Molecules* (G. N. LaMar, W. deW. Horrocks, Jr., and R. H. Holm, eds.), Academic Press, New York, 1973, p. 53.

3. G. N. LaMar, in *NMR of Paramagnetic Molecules* (G. N. LaMar, W. deW. Horrocks, Jr., and R. H. Holm, eds.), Academic Press, New York, 1973, p. 85.

4. W. deW. Horrocks, Jr., in *NMR of Paramagnetic Molecules* (G. N. LaMar, W. deW. Horrocks, Jr., and R. H. Holm, eds.), Academic Press, New York, 1973, p. 127.

5. R. A. Dwek, *NMR in Biochemistry*, Clarendon Press, Oxford, 1973, p. 174.

6. K. Wüthrich, *NMR in Biological Research: Peptides and Proteins*, Elsevier-North Holland, Amsterdam, 1976, p. 221.

7. G. N. LaMar and F. A. Walker, in *The Porphyrins* (D. Dolphin, ed.), Academic Press, New York, 1979, Vol. 4, p. 61.

8. G. Navon and G. Valensin, this volume, Chap. 1.

9. I. Bertini and C. Luchinat, this volume, Chap. 2.

10. J. D. Satterlee, this volume, Chap. 4.

11. H. M. McConnell and R. E. Robertson, *J. Chem. Phys.*, *29*, 1361 (1958).

12. R. J. Kurland and B. R. McGarvey, *J. Magn. Reson.*, *2*, 286 (1970).

13. E. Fermi, *Z. Phys.*, *60*, 320 (1930).

14. G. Williams, N. J. Clayden, G. R. Moore, and R. J. P. Williams, *J. Mol. Biol.*, *183*, 447 (1985).

15. W. deW. Horrocks, Jr., and E. S. Greenberg, *Biochim. Biophys. Acta.*, *322*, 38 (1973).

16. W. deW. Horrocks, Jr., and E. S. Greenberg, *Mol. Phys.*, *27*, 993 (1974).

17. C. D. Barry, H. A. O. Hill, P. J. Sadler, and R. J. P. Williams, *Proc. R. Soc. Lond.*, *A334*, 493 (1973).

18. B. Bleaney, *J. Magn. Reson.*, *8*, 91 (1972).

19. B. R. McGarvey, *J. Magn. Reson.*, *33*, 445 (1979).

20. R. G. Shulman, S. H. Glarum, and M. Karplus, *J. Mol. Biol.*, *57*, 93 (1971).

21. R. M. Keller, O. Groudinsky, and K. Wüthrich, *Biochim. Biophys. Acta.*, *427*, 497 (1976).

22. R. M. Keller and K. Wüthrich, *Biochim. Biophys. Acta.*, *533*, 195 (1978).

23. M. M. Maltempo and T. H. Moss, *Quart. Revs. Biophys.*, *9*, 181 (1976).

24. A. K. Gregson, *Inorg. Chem.*, *20*, 81 (1981).

25. S. J. Perkins, *Biological Magn. Reson.*, *4*, 193 (1982).

26. A. D. Buckingham and P. J. Stiles, *Mol. Phys.*, *24*, 99 (1972).

27. R. J. Abraham, G. R. Bedford, D. McNeillie, and B. Wright, *Org. Magn. Reson.*, *14*, 418 (1980).

28. R. J. Abraham, S. C. M. Fell, and K. M. Smith, *Org. Magn. Reson.*, *9*, 367 (1977).

29. R. J. Abraham, *J. Magn. Reson.*, *43*, 491 (1981).

30. O. D. Hensens, H. A. O. Hill, C. E. McClelland, and R. J. P. Williams, in *Vitamin B$_{12}$* (D. Dolphin, ed.), Academic Press, New York, 1982, p. 463.

31. H. Scheer and J. J. Katz, in *Porphyrins and Metalloporphyrins* (K. M. Smith, ed.), Elsevier, Amsterdam, 1975, p. 399.

32. J. L. Hoard, *Science, 174,* 1295 (1971).

33. W. R. Scheidt, *Acc. Chem. Res., 10,* 339 (1977).

34. W. R. Scheidt and C. A. Reed, *Chem. Rev., 81,* 543 (1981).

35. W. deW. Horrocks, Jr., and C. P. Wong, *J. Am. Chem. Soc., 98,* 7157 (1976).

36. T. Takano and R. E. Dickerson, *J. Mol. Biol., 153,* 95 (1981).

37. Y. Matsuura, T. Takano, and R. E. Dickerson, *J. Mol. Biol., 156,* 389 (1982).

38. E. Antonini and M. Brunori, *Hemoglobin and Myoglobin in Their Reactions with Ligands,* North-Holland, Amsterdam, 1971, p. 43.

39. T. Anderson, E. Thulin, and S. Forsén, *Biochemistry, 12,* 2487 (1979).

40. D. V. Behere, R. Birdy, and S. Mitra, *Inorg. Chem., 21,* 387 (1982).

41. C. M. Dobson and B. A. Levine, in *New Techniques in Biophysics and Cell Biology,* Vol. 3 (R. H. Pain and B. E. Smith, eds.), John Wiley, New York, 1976, p. 19.

42. F. Inagaki and T. Miyazawa, *Prog. in NMR Spectrosc., 14,* 67 (1981).

43. W. deW. Horrocks, Jr., and E. G. Hove, *J. Am. Chem. Soc., 100,* 4386 (1978).

44. G. N. LaMar, B. L. Budd, D. B. Viscio, K. M. Smith, and K. C. Langry, *Proc. Natl. Acad. Sci. USA, 75,* 5755 (1978).

45. G. N. LaMar, P. D. Burns, J. T. Jackson, K. M. Smith, K. C. Langry, and P. Strittmatter, *J. Biol. Chem., 256,* 6075 (1981).

46. M. Erecinska and J. M. Vanderkooi, *Meth. in Enzymol., 53,* 165 (1978).

47. W. deW. Horrocks, Jr., R. F. Venteicher, C. A. Spilburg, and B. L. Vallee, *Biochem. Biophys. Res. Commun., 64,* 317 (1975).

48. O. Jardetzky and G. C. K. Roberts, *NMR in Molecular Biology,* Academic Press, New York, 1981, p. 115.

49. G. R. Moore, R. G. Ratcliffe, and R. J. P. Williams, *Essays Biochem., 19,* 142 (1983).

50. R. M. Keller and K. Wüthrich, *Biochim. Biophys. Acta., 621,* 204 (1980).

51. F. S. Mathews, *Biochim. Biophys. Acta.*, *622*, 375 (1980).

52. G. R. Moore and R. J. P. Williams, *Eur. J. Biochem.*, *103*, 523 (1980).

53. G. R. Moore, R. J. P. Williams, J. C. W. Chien, and L. C. Dickinson, *J. Inorg. Biochem.*, *12*, 1 (1980).

54. L. M. Jackman and S. Sternhall, *Applications of NMR Spectroscopy in Organic Chemistry*, Pergamon, Oxford, 1969, p. 61.

55. N. J. Clayden and R. J. P. Williams, *J. Magn. Reson.*, *49*, 383 (1982).

56. M. R. Willcott III, R. E. Lenkinski, and R. E. Davis, *J. Am. Chem. Soc.*, *94*, 1742 (1972).

57. R. E. Davis and M. R. Willcott III, *J. Am. Chem. Soc.*, *94*, 1744 (1972).

58. W. C. Hamilton, *Statistics in Physical Science*, Ronald Press, New York, 1964, p. 157.

59. C. D. Barry, H. A. O. Hill, B. E. Mann, P. J. Sadler, and R. J. P. Williams, *J. Am. Chem. Soc.*, *95*, 4545 (1973).

60. H. A. O. Hill, P. J. Sadler, R. J. P. Williams, and C. D. Barry, *Ann. N.Y. Acad. Sci.*, *206*, 247 (1973).

61. S. Moreau, B. Perly, C. Chaehaty, and C. Deleuze, *Biochim. Biophys. Acta.*, *840*, 107 (1985).

62. R. J. P. Williams, *Structure and Bonding*, *50*, 79 (1982).

63. P. S. Burns, R. J. P. Williams, and P. E. Wright, *J. Chem. Soc. Chem. Commun.*, 795 (1975).

64. D. Lexa, J. Mispelter, and J. M. Saveant, *J. Am. Chem. Soc.*, *103*, 6806 (1981).

65. A. D. Boersma and H. M. Goff, *Inorg. Chem.*, *21*, 581 (1982).

66. J. E. Baldwin and P. Perlmutter, *Topics Current Chem.*, *121*, 181 (1984).

67. N. J. Clayden, G. R. Moore, R. J. P. Williams, J. E. Baldwin, and M. J. Crossley, *J. Chem. Soc. Perkin Trans. II*, 1693 (1982).

68. N. J. Clayden, G. R. Moore, R. J. P. Williams, J. E. Baldwin, and M. J. Crossley, *J. Chem. Soc. Perkin Trans. II*, 1863 (1983).

69. B. R. McGarvey, *Can. J. Chem.*, *53*, 2498 (1975).

70. R. G. Shulman, K. Wüthrich, T. Yamane, D. J. Patel, and W. E. Blumberg, *J. Mol. Biol.*, *53*, 143 (1970).

71. R. M. Keller and K. Wüthrich, *Biochim. Biophys. Acta.*, *285*, 326 (1972).

72. F. S. Mathews, M. Levine, and P. Argos, *J. Mol. Biol.*, *64*, 449 (1972).

73. R. Bois-Poltoratsky and A. Ehrenberg, *Eur. J. Biochem., 2,* 361 (1967).

74. C. Mailer and C. P. S. Taylor, *Can. J. Biochem., 85,* 1048 (1972).

75. H. Hori and H. Morimoto. *Biochim. Biophys. Acta., 200,* 581 (1970).

76. G. Williams, G. R. Moore, R. Porteous, M. N. Robinson, N. Soffe, and R. J. P. Williams, *J. Mol. Biol., 183,* 409 (1985).

77. G. R. Moore, M. N. Robinson, G. Williams, and R. J. P. Williams, *J. Mol. Biol., 183,* 429 (1985).

78. H. Senn, R. M. Keller, and K. Wüthrich, *Biochem. Biophys. Res. Commun., 92,* 1362 (1980).

79. M. P. Byrn, B. A. Katz, N. L. Keder, K. R. Levan, C. J. Magurany, K. M. Miller, J. W. Pritt, and C. E. Strouse, *J. Am. Chem. Soc., 105,* 4916 (1983).

80. R. J. P. Williams, G. R. Moore, and G. Williams, in *Prog. Biorg. Chem. Mol. Biol.* (Y. A. Ovchinikov, ed.), Elsevier, Amsterdam, 1984, p. 31.

81. G. Williams, G. R. Moore, and R. J. P. Williams, *Comments Inorg. Chem., 4,* 55 (1985).

82. D. L. Brautigan, B. A. Feinberg, B. M. Hoffman, E. Margoliash, J. Peisach, and W. E. Blumberg, *J. Biol. Chem., 252,* 574 (1977).

6

Relaxometry of Paramagnetic Ions in Tissue

Seymour H. Koenig and Rodney D. Brown III
IBM Thomas J. Watson Research Center
Yorktown Heights, New York 10598

1. GENERAL BACKGROUND

In somewhat over one decade, NMR imaging has progressed from a far-
sighted concept in the minds of an imaginative few to almost routine
clinical use, with over 200 whole-body imagers presently installed
in the United States alone. This new clinical modality is more than
a complement to x-ray computed tomography; it is the procedure of
choice in an increasing number of clinical situations, particularly
studies of the head and spine. The reason is its sensitivity to the
protons of tissue water and fat, in both soft tissue and bone marrow,
and its insensitivity to bone itself.

Proton density varies but slightly among tissues, certainly
not enough for images of proton density alone to be clinically useful.
Rather, the relaxation rates of tissue protons—in general, water
protons—vary significantly from one tissue to another, so that contrast in NMR images depends in the main on an appropriate weighting
of the signals from each picture element (pixel) of an image with
intensities that depend on proton density and the corresponding
relaxation rates. Technically, rate information is obtained by using
four radio frequency pulses grouped in closely spaced pairs such that
each pixel signal contains information about both T_1 and T_2. The
weighting by relaxation rates can then be manipulated during image
reconstruction to optimize the contrast for a given purpose.

NMR imaging has had a particular appeal as a noninvasive procedure, radiation-free and therefore ostensibly benign, in contrast
to x-rays, which carry an inherent risk. The static magnetic field
that pervades the patient appears to offer no intrinsic problems
(in the absence of pacemakers, internal surgical clips, shrapnel,
and the like; and carelessly placed steel tools), and induction
heating, from either the switching of the gradient coils (used for

producing a spatially varying resonant field) or the mandatory radio frequency pulses used to observe resonances, can readily be kept within conservative guidelines. However, given any image that has practical utility, there will always be the drive to improve its contrast and resolution, either to obtain a better image in a given situation or a comparable image in less time. Since contrast can be enhanced by altering relaxation rates in a tissue-specific way, it is no surprise that paramagnetic contrast-enhancing agents are becoming increasingly important in NMR imaging, despite the invasive nature of this approach.

There is a long tradition and a large body of accumulated experience with pharmaceuticals that transport chelated heavy-metal ions in vivo. These are used both for enhancing x-ray image contrast, particularly in regions accessible by body fluids, and for transporting radioactive nuclei to targeted tissues for radiographic images and therapy. Much is known of the toxicity and biodistribution of these agents, and it is no surprise that similar agents, substituted with paramagnetic ions from both the transition metal and lanthanide series, are being widely investigated as contrast-enhancing agents in NMR imaging and that some are already in clinical use in humans.

The fundamental limit to the utility of paramagnetic pharmaceuticals for enhancing contrast in NMR images will always be efficacy: the balance between contrast enhancement and the advantages gained, and the risks related to toxicity. Accordingly, it is important to know and understand the details of contrast enhancement (i.e., the relaxivity of these agents in vivo). This in turn relates to knowing both the physical and biochemical states of these agents in vivo, since both influence the relaxivities of these agents.

The measurement of the magnetic field dependence of nuclear (mainly proton) relaxation rates (NMRD profiles) of paramagnetic agents in (excised) tissue has been demonstrated to be a particularly powerful approach to the problem, in both its practical and fundamental aspects [1-3]. A significant amount of data has been accumu-

lated to date [4-6] which, based on the comparison of in vitro
measurements of tissue and water solutions containing the same
agents, makes clear that data from paramagnetic solutions can be
carried over rather directly to the analysis of enhanced relaxation
in tissue. In particular (and the temperature dependence of the
NMRD profiles of tissue is important in demonstrating this [6]),
tissue (to first order) can be regarded as a liquid, with tissue
water molecules able to sample the local intra- and intercellular
environments over a range of several cell diameters within a relaxation time [6,7]. Thus generally, tissue water can sample the macromolecular content of cells and the tissue content of paramagnetic
ions, and (as in solutions) generate proton relaxation rates that
are averages of a relatively microscopic environment.

Paramagnetic complexes that enhance the relaxivity of tissue
can be divided into two classes: endogenous and exogenous, i.e.,
those found or generated in vivo, and those deliberately introduced.
The endogenous agents are predominantly complexes of Fe^{3+} ions,
derived from oxidized hemoglobin, transferrin, and ferritin, but
can include O_2, free radicals, and perhaps Cu^{2+} in some disease
states. Exogenous agents in present clinical use tend to be strongly
chelated, low molecular weight complexes of Gd^{3+} [8], but NMRD profiles of binary and ternary complexes (with protein) of Mn^{2+} [2,9]
and Gd^{3+} [3,9] have been investigated, as have nitroxide radicals,
particularly those that are strongly lipophilic and therefore of
interest in relation to chemical shift imaging [10].

It is generally true, in the imaging range of 5-50 MHz (proton
Larmor frequency), that transverse relaxation rates ($1/T_2$) of tissue
are an order of magnitude greater than longitudinal rates ($1/T_1$) [11];
that the incremental contribution of contrast agents to $1/T_1$ and $1/T_2$
is about the same [4], so that the fractional changes in rates are
greater for $1/T_1$; and that the theoretical basis of relaxation by
paramagnetic centers is sufficiently understood so that the $1/T_2$ NMRD
profile can generally be predicted, with reasonable confidence, from
the $1/T_1$ profile [4]. This combined with the greater ease of measuring $1/T_1$ NMRD profiles is the rationale for restricting the remainder

of this work to a presentation and discussion of the $1/T_1$ NMRD
profiles of a range of paramagnetic complexes, both endogenous and
potential exogenous contrast-enhancing agents, all measured on samples of solutions and ex vivo tissue. Reference will be made in a
later section to the "Relaxometer"; for now we note that the instrumentation is highly specialized and designed for relaxometry (not
spectroscopy), and that comparable instrumentation exists in but a
few laboratories worldwide.

2. RELAXOMETRY OF Fe^{3+}

Iron is certainly the most ubiquitous endogenous paramagnetic ion.
However, only the high spin Fe^{3+} oxidation state, with its $S = 5/2$
configuration and relatively long electronic relaxation time, can
contribute significantly to proton relaxation [12-15]. The relaxivity
of Fe^{3+} complexes is quite complicated, due in part to the nature of
the aqueous chemistry of Fe^{3+}, in part to the (often) relatively slow
rates of ligand exchange, and in part to the tendency of the ground
electronic state of Fe^{3+} ions to split into levels in a manner that
makes prediction of relaxation effects difficult [16,17].

The potential clinical relevance of Fe^{3+} ions is multifaceted.
The management of hematomas (large coagulates of blood, as can accrue
from trauma) could in principle be aided by imaging the ions that
oxidize to Fe^{3+} as hemoglobin in the hematomas breaks down. The
increased production of ferritin and hemosiderin, proteins made in
response to excess iron that store it as Fe^{3+} in solid state cores
of a unique Fe^{3+} oxyhydroxide, can be monitored by NMR imaging [18].
Such excess iron is often the result of transfusion treatment of
patients who lack a gene for proper hemoglobin production. Death
can ensue when the storage mechanisms are ultimately overwhelmed.
As another example, transferrin is a circulating protein that trans-
ports iron, as Fe^{3+}, for ultimate incorporation in the bone marrow
into hemoglobin. Normally the two binding sites per molecule are
about 30% saturated with Fe^{3+}, an occupancy that can increase under

abnormal conditions. A question that arises is whether changes in
the iron loading of transferrin are detectable by imaging. As
another example, chelated forms of Fe^{3+} have potential utility as
contrast agents for the gut since mechanisms exist that prevent iron
uptake when it is not needed by the body. Finally, certain strongly
associated chelates of Fe^{3+} ions are taken up differentially by the
liver and therefore have potential as hepatobiliary contrast agents
[19].

An examination of the NMRD profiles of solutions of Fe^{3+} complexes, relevant to all the above, with discussions of the salient
features of the relaxation profiles, follows. The form of the theory
of relaxation appropriate to these (and other) systems, and the definition of the parameters that enter into a description of the NMRD
profiles, have been discussed in several chapters of this volume;
accordingly, the terms will be assumed to be familiar and will not
be defined unless there is a possibility of ambiguity.

2.1. Fe^{3+}-Aquoions

Since Fe^{3+} ions are insoluble above pH 3, under physiological conditions they will always be found complexed either to protein or small
chelates. This point is made explicit in the NMRD profiles [15] of
Figure 1. The details of the pH dependence involve equilibrium among
several hydrolyzed forms of aquoions [20], so that the exact pH dependence of the relaxivities at intermediate pH values depends on the
total Fe^{3+} concentration; however, Figure 1 is more than adequate to
illustrate its behavior.

In terms of a standard analysis, the NMRD profile arises, in
total, from an Fe^{3+}-proton magnetic dipolar interaction, with no
visible contribution from a contact interaction. The single dispersion, centered near 10 MHz, results from the condition $\omega_S \tau_c = 1$. The
relaxivity is mainly due to proton exchange between the bulk solvent
and the inner sphere of the Fe^{3+} ions, presumably acid-catalyzed
exchange of protons rather than exchange of water molecules. The

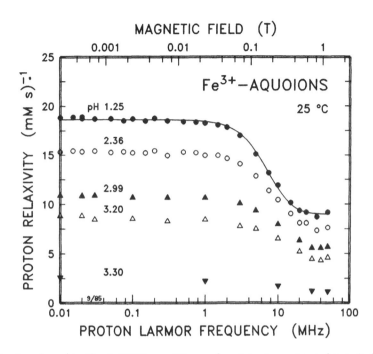

FIG. 1. Longitudinal NMRD profiles of solvent protons in a 0.68 mM solution of $Fe_2(SO_4)_3$, for several values of pH, at 25°C. The sample was initially at pH 1.25, and the pH was raised by addition of concentrated KOH. The data have been corrected for the effects of dilution, which were at most 7%. The solid curve through the upper set of data points results from a least squares comparison of the data with the usual theory of relaxation by dipolar interactions within a complex, with rapid ligand-solvent exchange. Outer sphere contributions (about 5%, cf. Fig. 2) were ignored in the fit. The values found for the several parameters of the theory are given in the text.

contribution from solvent molecules diffusing in the outer sphere environment of the aquoions is only about 5% of the total (see below) and will be ignored (but cf. Fig. 2). The correlation time τ_c is dominated by the rotational relaxation of the aquoion, but the fact that at high fields the profile does not decrease to 0.3 of its low field value, and perhaps even goes through a shallow minimum, indicates that a field-dependent τ_S makes a measurable contribution to τ_c. The solid curve through the upper data points results from a least squares comparison with the usual theory, using five adjustable parameters, and assuming six inner coordinated water molecules. The fit is insensitive to the residence time τ_M in that it is long enough (>1 nsec) not to contribute to the correlation time, and short enough (<<6 μsec) so that the rapid exchange limit can be assumed. The values for the other four parameters are r = 2.85 Å, τ_{So} = 84 psec, τ_V = 12 psec, and τ_R = 55 psec, certainly reasonable values in each case. (Note that the Stokes law relaxation time for a sphere of 3.6 Å radius in water, that of the hydrated aquoion, is 42 psec at 25°C.)

2.2 Small Chelate Complexes of Fe^{3+} Ions

Fe^{3+} ions can be maintained in solution at physiological pH by complexation with small chelates, either by using a large excess of a chelate that forms a weak complex, or by stoichiometric combination with a chelate that forms a tight complex [15]. Citrate is an example of the former. It is the dominant anion of certain dietary supplements meant to correct iron deficiency, and as such has been tried on human volunteers as a contrast agent for the gut [20]. Examples of the latter are DTPA (diethylenetriaminepentaacetic acid) [8] and EHPG (ethylenebis(2-hydroxyphenylglycine)) [19], which occupy the entire inner Fe^{3+} coordination sphere, and EDTA (ethylenediamine-tetraacetic acid), which tends to force Fe^{3+}, at least in part, into a seven-coordinate configuration [21]. The NMRD profiles of these chelates are shown in Figure 2 near physiological conditions of pH and temperature.

The excess relaxivity of the EDTA complex, compared with the other three, is evident, though the difference is not as great as one-sixth the inner sphere values (Fig. 1) nor equal to that of the isoelectronic Mn^{2+} (EDTA) complex, which has a value of 5 mM/sec at low fields, at 37°C. However, the Fe(EDTA) data (Fig. 2) do have a higher inflection field than the aquoion data (Fig. 1), higher than would be expected from the difference in temperatures at which

the profiles were taken. Correction of this would, at least in part,
make the intrinsic contribution of the coordinated water in Fe(EDTA)
closer to that of a single coordinated water of the aquoion. A more
quantitative analysis of the data would require knowing whether the
seven-coordinate bond length is that of the aquoion, or somewhat
longer, and whether or not there is a mixture of coordination configurations in solution.

The relaxivity profiles of the other three complexes (Fig. 2)
are ostensibly due to outer sphere effects, for which several pub-

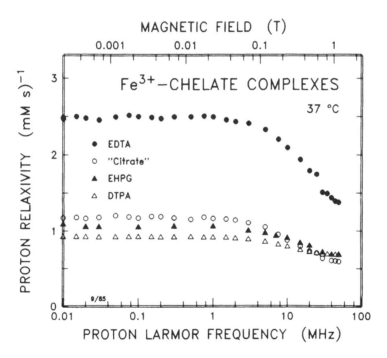

FIG. 2. NMRD profiles of solutions of several chelate complexes of
Fe^{3+}, near pH 7, at 37°C. EDTA (ethylenediaminetetraacetic acid) is
hexadentate, but the iron becomes (partially) seven-coordinate in
this complex; "citrate" here is the dietary supplement Geritol, in
which a large excess of ammonium citrate maintains Fe^{3+} in solution,
even near pH 7; EHPG [ethylenebis(2-hydroxyphenylglycine)] is a
hexadentate ligand taken up preferentially by liver; and DTPA
(diethylenetriaminetetraacetic acid) is an octadentate ligand.
The near equivalence of the lower three profiles suggests that outer
sphere relaxation dominates the relaxivity in all three cases.

lished theories [22-25] have been recently reconsidered [10]. Of
note in Figure 2 is that the high-field extreme of these profiles
does not disperse to the expected 0.3 of the low-field limit, indicating that not only does τ_S contribute to the fluctuations in local
field seen by water protons diffusing in the outer sphere environments
of the complexes, but that the field dependence of τ_S, characterized
by τ_V, must be included as well [cf. 4]. This requires the recent
generalization [10] of the results of Freed [24] to include τ_V.
However, application of the theory to the rather featureless data
for outer sphere relaxation (Fig. 2) is premature since we find that
a wide range of values for the several parameters of the theory fit
the data equally well, and there is as yet no realistic way of
deciding on their relative validities.

The major point to note is that Fe^{3+} can be made soluble near
physiological pH by small chelates, but at a cost of roughly a 10-
fold reduction in relaxivity values. The situation is only somewhat
improved by complexation with protein, if at all.

2.3. Fe^{3+}-Transferrin

Transferrin, a circulating serum protein of 84 kDa required for
transport of iron, has two iron-binding sites per protein molecule,
both very similar though not identical, and essentially noninteracting. The concentration of transferrin in the blood is ~0.04 mM,
and the molecules are about one-third saturated with Fe^{3+} under
normal conditions [26]. Though crystallographic data are not yet
available at high resolution, a large number of physical and biochemical investigations have led to the consensus that the Fe^{3+} ions
are six-coordinate, with four protein-donated ligands, and OH^- and
HCO_3^- contributed by solvent. HCO_3^- is required for tight binding of
Fe^{3+} and is always present in vivo unless specific mechanisms exist
to remove it.

Figure 3 shows the NMRD profiles of Fe^{3+}-saturated and apo-
transferrin [27,28], at 37°C. It is generally agreed that relaxation

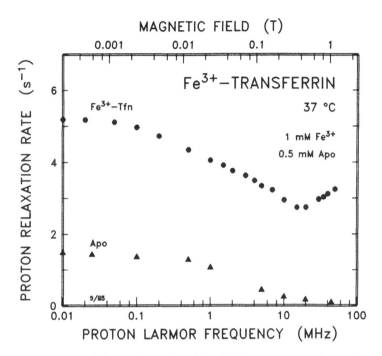

FIG. 3. NMRD profiles of an 0.5 mM solution of human transferrin saturated with Fe^{3+} (●), near pH 7, at 37°C, and that of the apo (demetallized) protein (▲).

results from rapid exchange of second-sphere, hydrogen-bonded water molecules, with a proton about 3 Å from an Fe^{3+} ion. (Exchange of an OH proton near physiological pH cannot be sufficiently rapid to contribute observably to relaxation, without resort to an unusual mechanism of exchange [28].) This type of second-sphere exchange, originally postulated for fluoromethemoglobin [14] and dubbed the "fluoromet mechanism" [28], also appears to account for relaxation in solutions of cupric and vanadyl transferrin derivatives [29], protein-bound nitroxide radicals [10], and perhaps Co^{2+}-substituted carbonic anhydrase (but cf. [30]).

The functional form of the contribution of the Fe^{3+} ions (Fig. 3) is rather complex, related to the rather anisotropic spin Hamiltonian of the Fe^{3+} spins in transferrin, and the many magnetic energy level crossings in the region of field comparable to the observed hyperfine and other low-field splittings [16,17]. What is germane for questions of imaging, however, is the magnitude of the paramagnetic relaxivity, which is in the range 3-4 mM/sec at all fields. Given the low concentration of transferrin in the circulation, saturation of this transferrin under iron overload conditions would contribute a rate increase of about 0.1 sec^{-1} to $1/T_1$ (and comparably to $1/T_2$), which would at best be on the edge of observability.

2.4. Fe^{3+}-Hemoglobin

The paramagnetic relaxivity of the Fe^{3+} ions of methemoglobin is comparable to, and in fact less than, that of Fe^{3+}-transferrin; the greater potential utility of hemoglobin in imaging comes about because of the greater concentration of hemoglobin relative to transferrin. Figure 4 shows the NMRD profiles of Fe^{3+}-hemoglobin and a diamagnetic control, carbon monoxyhemoglobin, at 35°C. The concentration chosen is the mean hemoglobin concentration of normal blood. The paramagnetic relaxivity (per Fe^{3+}) is very low, comparable to the outer sphere contribution of small chelates. It has previously been argued [14] that about half the methemoglobin relaxivity is outer sphere, with the remainder due to slow exchange of the inner coordinated water molecules of the heme groups. Indeed, displacing these waters by fluoride gives a 10-fold enhancement of the relaxivity due to second-sphere hydrogen-bonded waters, the "fluoromet mechanism" [28].

Oxidation of hemoglobin iron to Fe^{3+} at physiological concentrations will certainly produce a detectable change in an NMR T_1-weighted image, provided that access of water to the heme pocket is maintained by the structure of the physiological region in which this oxidation occurs. The incremental contribution to $1/T_2$ will be similar.

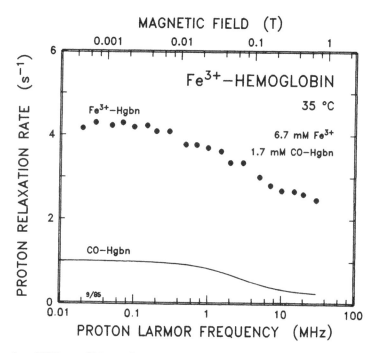

FIG. 4. NMRD profiles of a 1.7 mM solution of carbon monoxyhemo-globin (solid curve), which is diamagnetic, near pH 7, at 35°C, and a similar solution of aquomethemoglobin (●), which includes the diamagnetic background of the lower curve and the paramagnetic con-tribution of the four Fe^{3+} ions per protein molecule.

2.5. Ferritin

Ferritin, an iron storage protein of 447 kDa, has a spherical protein shell of about 130 Å outer diameter and 70 Å inner diameter assembled from 24 nearly identical subunits. Channels of both three- and four-fold symmetry lead to an inner core that, in the native state, may contain as many as 3,000 Fe^{3+} ions in a solid state, crystalline array of an unusual ferric oxyhydroxide that is ostensibly para-magnetic at physiological pH [31,32]. Figure 5 shows the NMRD pro-files for solutions of apoferritin, ferritin with one Fe^{3+} per sub-unit, and a ferritin with a rather fully loaded core [33]. The 24:1 data were obtained by titration of iron into apoferritin, as Fe^{2+}

with subsequent oxidation of the complexed iron to Fe^{3+} (the usual and necessary procedure). Preparation of the 900:1 sample was similar, but more complex in detail [33]. As might be expected (cf. Fig. 1 of [14]), loading the core makes little contribution to the $1/T_1$ relaxivity profile; the 30 Å thick protein shell isolates the solution from the core very effectively.

The surprise is that NMR images of patients with iron overload disease show dark livers and spleens [18], attributable to a very short T_2 caused by the presence of iron-induced ferritin formation, but at concentrations such that little if any effect would be expected on the basis of the data of Figure 5. The point is that the expectation, based on Figure 5, is that the $1/T_2$ profiles should be much like the $1/T_1$ profiles. This appears to be the case for the 24:1

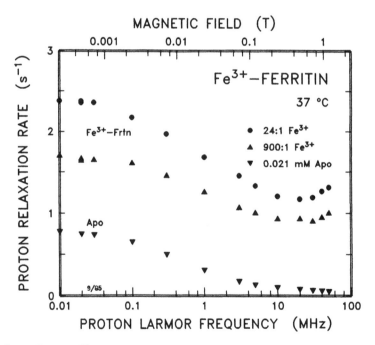

FIG. 5. NMRD profiles of a 21 µM solution of apo (demetallized) horse spleen ferritin (▼), near pH 7, at 37°C, and similar solutions with the ferritin loaded with respectively 24 (●) and about 900 (▲) Fe^{3+} ions per protein molecule.

sample, but $1/T_2$ for the 900:1 in vitro sample of a ferritin solution, and its value in vivo, is anomalously large, with a rate of
5 sec^{-1} at 20 MHz in vitro. No mechanism has yet been proposed that
accounts for this anomaly [33]. Even the effects of a Curie term in
the magnetization [34,35], known to produce an anomalous T_1/T_2 ratio
at very high fields (quadratic in field), do not account for the
observations, particularly the image data that indicate that the
high contrast observed in iron overload disease is rather independent
of the imaging field used and, in particular, is observed at relatively low fields [18].

2.6. Summary of Fe^{3+} Relaxometry

Ferric iron, to the extent that it will have a role in NMR imaging,
will be as an endogenous agent for contrast enhancement. In general,
Fe^{3+} complexes are poor relaxors, due in the main to the extensive
chelation needed to maintain Fe^{3+} in solution, with reasonable solvent
access, at physiological pH. Its preferred state is hydrolysis of
its coordinated water and precipitation [15,36]. The low relaxivities
of the complexes tend to be offset by the rather large concentrations
of Fe^{3+}, in methemoglobin, that can occur in disease or trauma.

A theoretical description of the NMRD profiles of Fe^{3+} complexes
tends to be rather difficult in general, in part due to the varied
(and unpredictable) biochemical mechanisms involved and in part to
the anisotropy of the ligand fields generally encountered.

3. RELAXOMETRY OF Mn^{2+}

Manganese is probably the most studied paramagnetic relaxation agent,
and for a variety of reasons. In the early years of applications of
NMR to solution chemistry, the discovery that the longitudinal and
transverse relaxivities of Mn^{2+} in aqueous solution were markedly
different [37] lead to the realization that a scalar part of the

Mn^{2+}-proton interaction in the aquoion produced a relaxivity contribution as large as the magnetic dipolar interaction [38] and to the possibility of using this interaction to probe the nature of the ligand orbitals. Subsequently, when interest developed in applying relaxometry to studies of macromolecular systems, the fact that Mn^{2+} could substitute for Mg^{2+} in many loose macromolecular complexes made possible studies of the kinases [39], including their interactions with substrate and ATP, as well as studies of various nucleic acids [40-42]. Mn^{2+} also has the desirable properties that it is soluble at physiological pH; it has an EPR spectrum that is readily observable and identifiable in solutions of the aquoion and some protein complexes; and it can often substitute for diamagnetic ions in many metalloproteins without altering substantially certain important properties of the native proteins.

Further developments in relaxometry of Mn^{2+} showed that the NMRD profiles of its many complexes are quite specific [43]: generally, based solely on the functional form of the profiles, aquoions, small chelate complexes, and macromolecular complexes can all be distinguished in a rather unique fashion [44].

3.1. Range of NMRD Profiles of Mn^{2+}

Figure 6 shows the NMRD profiles of solutions of three complexes of Mn^{2+}: the hexaaquoion, Mn^{2+} complexed tightly with excess EDTA, and Mn^{2+}-concanavalin A, a well-studied lectin that contains Mn^{2+} in the natural state. The reasons for the different profiles are well understood and only the general aspects will be considered here [cf. 4]. For the aquoion, the low-field dispersion, near 0.1 MHz, arises from the contact interaction and inflects when $\omega_S \tau_S \sim 1$, corresponding to τ_S of about 2.5 nsec. The exchange time of a water ligand is known to be about 20-fold longer, and therefore does not influence the correlation time, nor does the rotation of the complex (by definition of a scalar or contact interaction). The high-field dispersion, near 10 MHz, corresponds to the condition $\omega_S \tau_c \sim 1$,

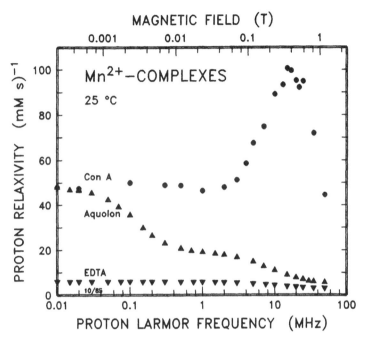

FIG. 6. NMRD profiles of solutions of three different complexes of
Mn^{2+}: the protein Mn^{2+}-concanavalin (●); the aquoion (▲); and the
EDTA complex (▼), all near pH 6.4, at 25°C.

where τ_c is clearly about 25 psec and dominated by the rotational
relaxation of the aquoion.

The contact interaction is about 1% of the dipolar interaction;
its 100-fold longer correlation time is what makes its contribution
to the NMRD profile comparable to that of the dipolar interaction.
When Mn^{2+} is complexed with a macromolecule, as in Figure 6, with a
relatively long rotational relaxation time of about 60 nsec for the
complex, the correlation times for both the contact and dipolar
interactions become about the same; both are dominated by τ_S, and
no contribution from the contact interaction is seen. However, a
new aspect appears in the NMRD profile [43], which is the peak, near
15 MHz for this protein. It arises because τ_S is itself field-
dependent [40,41,43], beginning to increase at a field determined

by the condition $\omega_S\tau_V \sim 1$. Here τ_V characterizes the electron-
protein ("lattice") interaction and, for macromolecular systems,
its source and relatively long value of about 0.1 nsec, remains
a mystery.

A dispersive component due to a contact interaction is also
absent in the profile for the Mn(EDTA) complex. The reason is not
clear, since it is known that Mn^{2+} becomes seven-coordinate when
complexed with EDTA [45] and that about half the resulting relaxivity
arises from exchange of a water from the open coordination position
of the Mn^{2+} ion [5]. The EPR spectrum is known to broaden upon
chelation [45], and this shortened electronic relaxation time is
the most likely reason for the absence of a contact term in the
Mn(EDTA) profile.

3.2. Outer Sphere Relaxation of Chelated Mn^{2+}

It is possible to make small chelate complexes of Mn^{2+} for which all
the relation is due only to outer sphere interactions. NOTA, a
triazacyclononane with three acetate groups coordinated to the
nitrogens, is a highly symmetric hexadentate molecule that forms
a relatively rigid chelate complex with Mn^{2+}, which remains six-
coordinate. DOTA, the analogous four-nitrogen molecule, is octadentate. Solutions of complexes of these chelates with Mn^{2+}, as
well as Mn(DTPA), have nearly identical NMRD profiles [46], as seen
in Figure 7. The relaxivities are about half those of Mn(EDTA).

The solid curve through the profile for Mn(DTPA) (Fig. 7)
results from a least squares comparison of the data with a theory
of outer sphere relaxation [10] which, as noted above, involves
four parameters. Unfortunately, but not surprisingly, the solid
curve is a representation of two sets of values for these parameters,
both of which give the identical fit over the field range illustrated.
We find the following pairs of values: the distance of closest
approach a, 3.48 and 3.21 Å; the relative diffusion constant D,
2.77 and 2.84 x 10^5 cm^2/sec^{-1}; the low-field value of the electronic
relaxation τ_{So}, 231 and 137 psec; and τ_V, 39 and 5.5 psec.

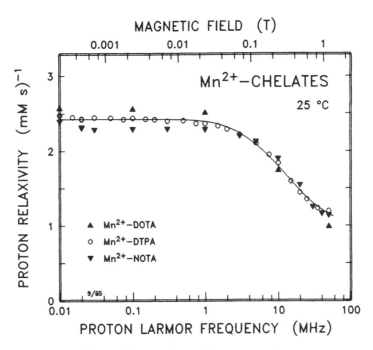

FIG. 7. NMRD profiles of solutions of several chelate complexes of
Mn^{2+}, near pH 6.5, at 25°C. NOTA is hexadentate, whereas DTPA and
DOTA are both octadentate. The near equivalence of the three profiles suggests that outer sphere relaxation dominates the relaxivity
in all three cases. The solid curve through the Mn(DTPA) data points
(o) results from a least squares comparison of the data with the
theory of outer sphere relaxation. The values found for the several
parameters of the theory are given in the text.

     At present we have no reason to prefer one set of values over

the other, nor are we certain that there aren't other equally good

values for the parameters. One might argue in favor of the set with

τ_V = 5.5 psec since this is close to the value found for the Mn^{2+}-

aquoion. The latter is established mainly from analysis of the

field dependence of $1/T_2$ at high fields, which is dominated by the

contact interaction. However, this interaction is not seen in outer

sphere relaxation, so that there is no simple check on the validity

of this option.

248   KOENIG AND BROWN

3.3. Mn^{2+} in Tissue

Mn^{2+} is an exogenous contrast agent, and there is some evidence that
it can accumulate in the liver of rabbits and contribute observably
to the liver relaxation rate when the rabbit's food intake is restricted to certain prepared chows. The experience to date with
externally introduced Mn^{2+}, either given by intravenous injection
or by mouth, chelated or as the chloride, is that it accumulates
throughout the liver, separated from chelate (if any) and complexed
with some macromolecular structure [2]. This is illustrated in
Figure 8, which compares the NMRD profiles of (excised) livers of
normal rabbits before and after intravenous injection of a weak

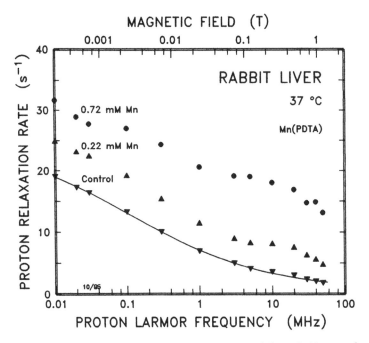

FIG. 8. NMRD profiles of normal rabbit liver (▼) and livers from
two rabbits injected with different total amounts of Mn(PDTA)
(propanoldiaminetetraacetic acid) at 37°C. Rabbits were sacrificed
15 min after injection. The Mn contents of the livers were subsequently determined to be 0.72 mM (●) and 0.22 mM (▲) for the two
cases.

chelate complex of Mn^{2+}. The injections were 12 and 36 µmoles per kg body weight of Mn^{2+}-propanoldiaminetetraacetic acid (PDTA), and the injected animals were sacrificed 15 min after injection. The mean concentration of Mn in the livers was measured by ICP (inductively coupled plasma) analysis of the actual samples used; the findings are indicated in the figure.

The lowest set of data, the control profile, is from an uninjected rabbit. Runs on numerous samples of such livers show little individual variation, except for the slight rise in the 10- to 20-MHz region. This rise, attributable to Mn^{2+} in the food, has been ignored in fitting the solid curve through these data, a curve used later to correct the profiles of the Mn-containing livers for the diamagnetic background. There are three points to make: the concentration of Mn found in the liver is proportional to the total amount injected in the limited range shown [47]; there is a marked enhancement of the relaxation rates; and this enhancement is not linear in Mn concentration for the examples used here. In fact, we know from earlier work that there is indeed a saturation in the relaxation rates with Mn(PDTA) dosage, and from more recent work [47] that this saturation arises from chemical saturation of the liver with Mn at about 0.5- 0.7 mM Mn. Indeed, when this regime is approached, the relaxation data become nonexponential and the entire situation more complex [47]. Accordingly, we consider only the data for the lower Mn(PDTA) dosage, replotted in Figure 9 as the paramagnetic contribution to the liver NMRD profile per mM of Mn.

The data in Figure 9 have a clear peak, indicative of immobilized yet solvent-accessible Mn^{2+}. The relaxivities are high, at least 10-fold greater than the outer sphere relaxivities of small Mn^{2+}-chelate complexes. Based in part on the fact that administration of aqueous Mn^{2+} gives the same results, we argue that the data (Fig. 9) represent macromolecular bound Mn^{2+} ions freed from chelate [2]. It is not yet clear where these ions are bound. Addition of Mn^{2+} ions to a suspension of liver cells, removed enzymatically from their tissue substrate, gives very similar NMRD profiles immediately after Mn^{2+} addition [6], suggesting that the Mn^{2+} may be bound to the

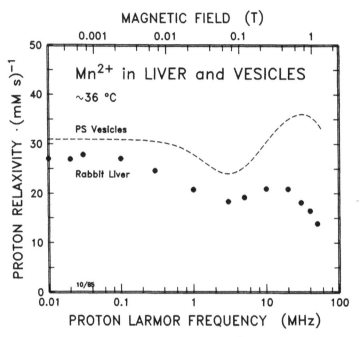

FIG. 9. The paramagnetic contribution of Mn^{2+} to the NMRD profile
of the liver of Figure 8 (●), containing 0.22 mM Mn, indicated in
relaxivity units. The dashed curve is the profile of a solution
of Mn^{2+} bound to phosphatidylserine vesicles (Kurland and Koenig,
unpublished), near 36°C.

exterior of the cells, associated with the cell membranes. On the
other hand, in vivo data indicate that the Mn^{2+} accumulates in the
bile, which can only occur if it passes through cells of the liver.
Though this is not of concern at the moment, one should note that
physiological questions of this sort can be addressed by NMRD measurements, perhaps more readily than by any other technique.

The dashed curve in Figure 9 is the dispersion profile of Mn^{2+}
ions added to a solution of phosphatidylserine vesicles [2,48], the
intent being to compare the liver data with a system that mimics cell
membranes. The exact form and amplitude of the vesicle profiles are
sensitive to the ionic strength and cationic content of the solution;
nonetheless, the qualitative similarity of the two profiles (Fig. 9)

is suggestive of a generalized interaction of Mn^{2+} ions in liver
with the exterior polar groups of the cell membranes.

3.4. Remarks on Mn^{2+} Relaxometry

The finding that chelated Mn^{2+} becomes dissociated in the liver is
a strong argument for looking for other paramagnetic contrast agents
useful for clinical NMR imaging. Current interest centers around
chelates of Gd^{3+} [8], which tends to be nine-coordinate, so that
chelation by even an octadentate ligand leaves an inner sphere site
for solvent exchange (see below). For the present, because of the
toxicity of Mn^{2+}, its utility will be restricted to animal experiments of the type indicated here, centered around liver function
and chemistry, bile production, and gall bladder activity. The unique
relaxometry properties of Mn^{2+}, as summarized in Figure 6, make it
rather clear that such interest will continue.

4. RELAXOMETRY OF Gd^{3+}

The trivalent gadolinium ion has a half-filled f shell and, accordingly, a septet S ground state with a relatively long electronic
relaxation time [cf. 49] (though generally not as long as Mn^{2+}).
The ion is large and highly charged, so that the coordination number
tends to be high, typically nine, compared with six for Mn^{2+}. Hexadentate chelates exist that form very strong complexes with Gd^{3+},
with three coordinated waters in aqueous solution. More popular at
present for clinical imaging is the complex of Gd^{3+} with the octadentate chelate DTPA [8]. It is small, highly stable, very soluble
under physiological conditions, and excreted in the urine within
several hours after intravenous introduction [3,8]. Its toxicity
is low, as is its relaxivity in aqueous solution [5], though the
latter is greater than the comparable Mn^{2+} complexes because of both
a higher spin (7/2 vs. 5/2) and a larger inner sphere coordination
capacity.

The applications of Gd(DTPA) to date have been to the examination of breaks in the blood-brain barrier (e.g., due to stroke); the relatively low and comparable values of $1/T_1$ and $1/T_2$ for blood, and body fluids in general, favor the utility of paramagnetic contrast agents, which generally make comparable contributions to the *changes* in $1/T_1$ and $1/T_2$. (In the imaging field range, one generally finds $T_2 \ll T_1$ for soft tissue; image data are processed by T_2 weighting, to minimize the time required for obtaining a useful image.) However, little is known about the interaction of Gd(DTPA) with tissue (including blood), and to what extent its in vivo relaxivity relates to the values measured in solution. Moreover, it has only recently been demonstrated that Gd(DTPA) remains chelated in vivo (or at least in the kidney) and is excreted in this chemical form [3]. The relevant background data follow.

4.1. Range of NMRD Profiles of Gd^{3+}

Figure 10 shows the NMRD profiles of solutions of four complexes of Gd^{3+}: the aquoion, Gd(EDTA), Gd(DTPA), and Gd^{3+}-concanavalin A, in the latter case bound at a site different from the Mn^{2+}-binding site relevant to the profile (Fig. 6). In many ways the behavior of Gd^{3+} is much like that of Mn^{2+}, particularly when allowance is made for the difference in coordination number. The main distinction between the results in Figures 6 and 10 is in the profiles of the aquoions: Gd^{3+}, because its f shell is not part of the ligand environment as is the d shell of Mn^{2+}, has no detectable contact contribution. As a result, the profiles of the chelates of Gd^{3+} and the aquoions are very similar in form, generally making it impossible to tell from the NMRD profile whether Gd^{3+} in a sample exists as the aquoion or in a small chelate; the total Gd^{3+} content must be known from other measurements. This is quite unlike the situation for Mn^{2+}, for which the size of the contact contribution is a quantitative measure of the concentration of Mn^{2+} aquoions [44].

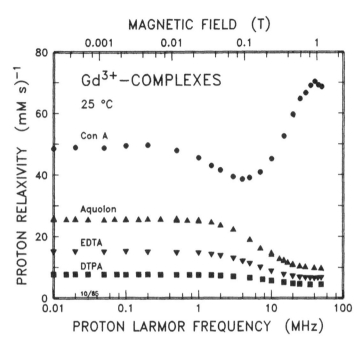

FIG. 10. NMRD profiles of solutions of four different complexes of Gd^{3+}: the protein concanavalin A, with the native Mn^{2+} replaced by Zn^{2+} and the Gd^{3+} bound nonspecifically ($\bullet$); the aquoion ($\blacktriangle$); the EDTA complex ($\blacktriangledown$); and the DTPA complex ($\blacksquare$), all near pH 6.4, at 25°C.

Another distinction between the properties of Gd^{3+} and Mn^{2+} that influences the NMRD profiles of small complexes of Gd^{3+} and τ_S of Gd^{3+} is comparable to the rotational relaxation time of these complexes, whereas for Mn^{2+}, τ_S is typically about 10-fold longer. Moreover, τ_S for Gd^{3+} (and Mn^{2+}) generally becomes field-dependent near the dispersive regions of the NMRD profiles. Consequently, the functional form of the profiles, and their dependence temperature, could also arise from and be confused with incipient slow exchange [49,50]. In general, it is difficult to distinguish between the two conditions without resort to other information, such as EPR data or NMRD profiles of, say, deuterons, in comparable samples [50]. This situation was considered in some detail previously [cf. 49,50].

4.2. Unchelated Gd^{3+} in Tissue

There are as yet no data for Gd^{3+} comparable to those of Figure 8 for Mn^{2+}, i.e., no instance in which the presence of protein-immobilized Gd^{3+} in tissue is evident from a high-field peak in the NMRD profile. There are several reasons for this, among them that chelated Gd^{3+} concentrates in the kidney medulla, unlike chelated Mn^{2+}, which dissociates and concentrates in the liver. As will be illustrated below, the Gd^{3+} chelate complexes remain rotationally mobile in the medulla. On the other hand, if Gd^{3+} is injected into the blood as the aquoion, it "disappears," i.e., it becomes relatively "relaxation-silent," this despite the fact that serum albumin is known to enhance the relaxivity of Gd^{3+} in water or buffer solutions [51]. One reason, only recently clarified [52], is that Gd^{3+} can interact with the normal levels of bicarbonate in blood and (presumably) precipitate as the carbonate. Thus, to take advantage of the high relaxivities of Gd^{3+} macromolecular complexes for in vivo studies will require an approach other than those tried to date, perhaps the use of ternary Gd^{3+}-protein-chelate complexes (see below).

4.3. Small Chelate Complexes of Gd^{3+}

The relaxivity of small chelate complexes of Gd^{3+} is turning out to be rather intriguing; the results to date suggest that the meaning of outer sphere relaxation be carefully reexamined and, in particular, that careful attention be given to the possibility of hydrogen bonding of water molecules in the second coordination environment of the chelated Gd^{3+} ions, with resident lifetimes sufficiently long to influence the relaxivity [29,53]. A hint of this phenomenon comes from a comparison of the NMRD profiles of Gd(EDTA) and the aquoions (Fig. 10). EDTA, being hexadentate, should remove two-thirds of the waters coordinated in the aquoions. Since the correlation frequencies for the two profiles are about the same (a fit gives 6.3 MHz for the

aquoion and 5.9 MHz for the chelate complex), the low-field relaxivity of Gd(EDTA) might be expected to be about 8 mM/sec; the observed
value is two-fold greater. The paradox is more apparent in the next
figure.

Figure 11 compares the NMRD profiles of Gd(DOTA) and Gd(NOTA),
octa- and hexacoordinate chelates, respectively [46]. Over the entire
field range, Gd(DOTA), which ostensibly has one-third the number of
coordinated waters of Gd(NOTA), nonetheless has the higher relaxivity.
Moreover, the relaxivities of both are greater than those of Gd(DTPA),
also octadentate. Though the variation in profiles among these several chelate complexes can in part be ascribed to differences in τ_S
(good EPR data are needed), it does appear that some of the chelate

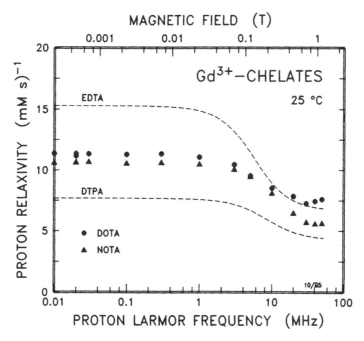

FIG. 11. NMRD profiles of Gd(DOTA) (●) and Gd(NOTA) (▲), near pH
6.4, at 25°C. The relaxivity of the former, with ostensibly one-
third the number of coordinated water molecules of the latter, is
nevertheless greater at all fields. The dashed curves indicate the
profiles of Gd(EDTA) and Gd(DOTA), for ease of comparison.

complexes have an excess relaxivity over and beyond what can be
accounted for by inner coordination plus the traditional outer
sphere contributions. This excess may be another example of the
fluoromet mechanism [14,28].

Displacement of the Fe^{3+}-coordinated water in methemoglobin
by F^- increases the paramagnetic part of the relaxivity (Fig. 4)
about 10-fold. This was attributed to hydrogen-bonded, second-sphere
waters, the F^--coordinated protons being not much farther from the
paramagnetic ions than the protons of coordinated first-sphere water
molecules. Moreover, the labile hydrogen bond allows for rapid
exchange, but apparently not so rapid as to shorten the correlation
time appreciably and therefore reduce the relaxivity contribution
to something unobservable. This fluoromet mechanism has been invoked
to account for the relaxivity profiles of Fe^{3+}-transferrin [28] (Fig.
3) as well as those of Cu^{2+}- and VO^{2+}-transferrin [29]. More recently,
it was suggested as the main contribution to the relaxivity of protein-
bound nitroxide radicals [10]. The values for τ_M in all these cases
would appear to be no shorter than about 10 nsec. The question now
is whether water molecules hydrogen-bonded to those carboxylate oxy-
gens coordinated directly to chelated Gd^{3+} ions, for example, can
have sufficiently long residence times so that their contributions
must be considered separately from the usual outer sphere contributions. This is an open question, at present, and an extremely important one for NMR imaging, in which the optimal balance between toxicity and relaxivity invariably is sought.

4.4. Gd^{3+} in Tissue

Gd^{3+} is an exogenous contrast agent, at present used widely in human
clinical medicine as the Gd(DTPA) complex [8]. Not surprisingly, the
majority of the investigations of the relaxivity of Gd^{3+} complexes in
vivo have involved this particular complex. Having been removed from
the circulation by the renal (kidney) cortex, Gd(DTPA) accumulates in
the renal medulla where it is concentrated by active countercurrent
processes for ultimate excretion in the urine.

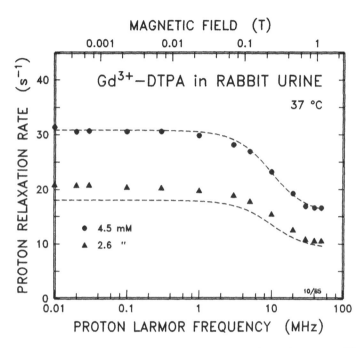

FIG. 12. NMRD profiles of two samples of rabbit urine, at 37°C,
found to contain, respectively, 4.5 mM (●) and 2.6 mM (▲) total Gd.
The dashed curves associated with the data points are the profiles
expected for solutions of the same respective concentrations of
Gd(DTPA) in water, at physiological pH and 37°C.

Preliminary results for rabbit urine [3] are shown in Figure
12. Here, the observed profiles are compared with those expected
for solutions of pure water containing Gd(DTPA) equal in concentration to the measured total Gd content of the samples. The agreement
is extremely good, indicating within the experimental uncertainty
that Gd^{3+} remains chelated in its passage through the rabbit. Complementary results can be obtained by radioactive tracer techniques,
for example, but these only determine total Gd; few techniques other
than NMRD can demonstrate the chemical state of the contrast agent
so readily.

Demonstration of the chemical state of Gd in the medulla itself
is more difficult, however, due in part to the function-correlated
structure of this organ. Since its function is to concentrate solute

ions for excretion, it therefore must have compartments among which
the rate of water exchange can be controlled. If, in addition, the
distribution of Gd(DTPA) is also spatially nonuniform, with differential access by the tissue water, then the time dependence of the
magnetization data that lead to each data point of an NMRD profile
becomes subtly nonexponential.

An attempt to resolve each point of an NMRD profile of tissue
into two contributing relaxation rates has only recently been successful [3]. There are several intrinsic difficulties: the data set which
is to be resolved into two exponentials can generally be described
quite adequately by a single exponential, with an uncertainty of less
than ±3%; for reasonable values of the signal-to-noise ratio, it is
almost a mathematical truism that two exponentials can often be simu-
lated by one; accordingly, a large data set must be accumulated with
rather high accuracy. An example is shown in Figure 13, where over
100 points are shown along the magnetization decay curve of a sample
of excised renal medulla for a single field. (For the profiles in
the previous figures, 23 points were taken for each point on the
profiles, and the fit to a single exponential was invariably better
than ±1%.) The results are interesting in their own right in that
it will be necessary to include the anatomy and physiology of the
renal medulla in a very specific way in order to model the data.
However, more information is needed to prove that the Gd in the
medulla is present as Gd(DTPA); an additional approach is necessary,
one in which the tissue structure is partially disrupted to allow
tissue water uniform access to the exogenous Gd.

Figure 14 shows data for three samples of rabbit renal medullas
with differing measured concentrations of Gd. Two samples were aged
(for 11 weeks) at 5°C to permit a certain amount of enzymatic degradation and, perhaps, extensive diffusion of water and contrast agent.
A control, with no Gd, was treated similarly. A third sample, fresh,
was cut into millimeter size pieces and subjected to repeated cycles
of freezing and thawing. The profiles of all four samples were
measured in the usual way to obtain the single experimental NMRD

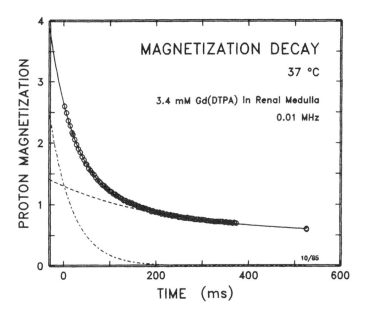

FIG. 13. An example of the time dependence of the proton magnetization of a sample of rabbit renal medulla containing about 3.4 mM
Gd(DTPA) that is used to compute a single typical point of an NMRD
profile, here 0.01 MHz, at 37°C. Because of the high load of Gd
in this sample, the data were measurably nonexponential. A fit of
a single exponential to the magnetization decay gives a rate of
12.1 sec^{-1} with a standard deviation of 3.6%, whereas a fit of two
exponentials gives rates of 20.1 sec^{-1} and 4.2 sec^{-1} with an uncertainty of 0.76%. The relative amplitudes of the two fits are about
3:1, obtained by extrapolating the data back to -30 msec, a reasonable approximation for the instrument settings used. About 100
points were taken along the decay curve, including eight at the
long-time extreme.

profiles. In every case, the data were well described by a single

exponential within the experimental uncertainty. The profile of

the aged control was indistinguishable from that of fresh tissue

and that for the lowest dosage was also little changed. However,

the amplitudes of the profile of the intermediate dosage sample

increased by 50% and that of the most heavily dosed sample by 100%

from the respective single exponential profiles of comparable fresh,

intact samples.

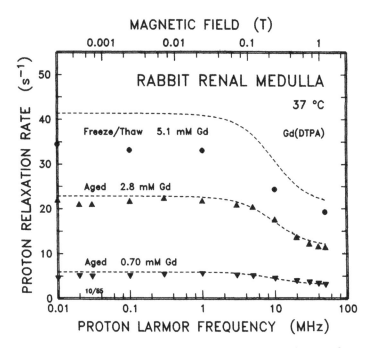

FIG. 14. Paramagnetic contribution to the NMRD profiles of three
samples of rabbit renal medulla, at 37°C, found to contain, respectively, 5.1 mM (●), 2.8 mM (▲), and 0.70 mM (▼) total Gd. The data
are for samples in which the tissue structure has been partially
degraded, either by several cycles of freezing and thawing or by
autocatalysis due to long-term (11 weeks) storage at 5°C. The magnitude of the upper profile increased about 100%, the center one
about 50%, and the lower one hardly at all as a result of these
procedures. The data after this treatment could all be well represented by single exponentials. The curves associated with the data
points are the profiles expected for solutions of the same respective concentrations of Gd(DTPA) in water, at physiological pH and
37°C, but including a correction for the volume occupied by the
solid content of tissue (about 20 wt%).

The profiles (Fig. 14) compare well with the profiles expected

from water solutions of Gd(DTPA) at concentrations of Gd found in the

tissue samples, as shown by the smooth curves. (The curves are ad-

justed to take account of the 20 wt% solids content of the tissue,

which makes the Gd concentration in tissue water greater than its

measured mean value.) The comparison of the curves with the data

for the aged samples is excellent, this being the strongest evidence
to date that not only is Gd present in the renal medulla as Gd(DTPA),
but the microviscosity sensed by the Gd(DTPA) in all the water in
the tissue is essentially that of neat water at the same temperature.
For the freeze/thaw sample the agreement is quite reasonable and
would probably be improved by further disruption of the tissue structure. In all, the main point would appear proven; certainly the
procedures necessary have become clear.

5. PRESENT AND FUTURE

5.1. The Nature of Tissue

Throughout the preceding discussion it has been assumed implicitly
that the results obtained in vitro for homogeneous solutions of paramagnetic complexes can be carried over (with little change) to considerations of the same complexes in tissue. Given the complexities
of tissue, both structural and compositional, the validity of such
an approach may appear questionable at best. However, the phenome-
nology demonstrates otherwise. One finds that, at least to first
order, water of tissue is much like solvent water in homogeneous
solutions of macromolecules with comparable solute concentrations
(~20%). This is the rule, with specialized organs like the renal
medulla being the exceptions.

Empirically, it is found that water in tissue is liquid and
highly mobile. The structure of tissue, much like the structure of
some gels, serves to confine the water without altering its gross
dynamics. An example is found in the comparison of the NMRD profiles
of congealed blood and fresh blood [7], which are indistinguishable
in form, magnitude, and temperature dependence. Moreover, the temperature dependence of the NMRD profiles of both blood and more complex tissues [6] shows that the application of the concepts of motional
narrowing, the essence of liquid behavior, is appropriate; rates decrease as temperatures increase. Additionally, the relatively low
longitudinal relaxation rates of diamagnetic tissue at high fields,

coupled with the intensity of the signal (which indicates that
essentially all tissue water contributes), shows that the microscopic viscosity of tissue water, i.e., its microscopic Brownian
motion, cannot be grossly different from that of pure water and
solvent water in protein solutions. Moreover, the paramagnetic
contribution of small chelate complexes to the profiles of various
tissues, exemplified by the results (Fig. 13), supports this view;
an increased viscosity would increase the correlation time, and
thereby lower the inflection field and increase the low-field
relaxivity of the profile shown. This is not seen here, nor in a
variety of other instances (unpublished observations).

The advent of NMR imaging has focused attention on the importance and the utility of investigating applications of NMRD techniques to *biological* systems. The facility with which this has
been possible to date relates to the long experience, gained in
many laboratories, with analogous applications to *biochemical*
systems, much of which is the subject of other chapters in this
volume. It becomes reasonable to ask, then, what new directions
in magnetic resonance-related areas will be stimulated by the
rapidly evolving field of NMR imaging. Several of these are considered below.

5.2. New Directions

5.2.1. *Transverse Relaxation Rate Profiles*

Current understanding of relaxation processes in homogeneous solutions of noninteracting paramagnetic solute complexes is sufficiently
advanced that one can generally predict the transverse ($1/T_2$) NMRD
profile from the longitudinal ($1/T_1$) profile [cf. 4]. This is particularly convenient since the $1/T_2$ profile is considerably more
difficult and time consuming to measure [43]. Moreover, for the
paramagnetic contribution of independent ions, the two profiles are
rather similar unless there is a significant contact contribution
to $1/T_1$ at low fields [cf. 4]. These generalities also hold for

the more limited data for tissue. However, in tissue, there is a
good likelihood that other situations will arise because of fine-
scaled magnetic inhomogeneities of the samples induced by small
ferromagnetic particles.

Such particles, typically within an order of magnitude of
10^3 Å in diameter, are ingested by macrophage cells specialized to
ingest and destroy viruses and bacteria. The delineation of tissue
structure under these circumstances can be particularly useful in
certain clinical situations, and experience to date using such particles as exogenous contrast agents indicates that the influence on
the observed $1/T_2$ cannot be readily predicted from $1/T_1$. In reality,
such particles introduce large, but localized, magnetic gradients
that contribute to precessional phase loss of the tissue protons,
often called $1/T_2^*$; the net effect on imaging is as though $1/T_2$ of
tissue has been increased substantially. Though little work on
homogeneous solutions or suspensions of such particles has been
reported to date, there is every reason to anticipate that in vivo
behavior can be readily related to in vitro solution behavior. What
will be needed is improved instrumentation and experimental protocols
for obtaining transverse NMRD profiles rapidly and accurately.

5.2.2. Nonaqueous Systems

Improvements in magnet technology, particularly the achievement of
good homogeneity over large volumes at high fields, has made possible
high-resolution images at 80 MHz (2 Tesla), with the promise of suc-
cess at much higher fields. A qualitative change appears here,
namely, a spatial displacement of adipose (fat) tissue in the NMR
images due to the chemical shift of the methylene protons of fat
compared with water protons. This has stimulated work on the relaxation properties of neat fatty acids as models for body fat and lipophilic paramagnetic entities that may serve as contrast agents specific for adipose tissue [10]. Among these are nitroxide radicals
which, though they have relatively low relaxivities in water solution,
develop particularly interesting properties in lipids and/or when

complexed with macromolecules. Research on these systems and on the
related ternary systems involving water solutions of fatty acids,
vesicles, and paramagnetic ions will find application to the more
complex biological systems studied by NMR imaging.

5.2.3. Other Nuclei

Protons are not the only endogenous nuclei that can be used for
producing NMR images of humans and animals; sodium is another, and
phosphorus could conceivably be a third. The relaxometry of sodium
is complicated by the electric quadrupole moment of the sodium
nuclei; the spin of 3/2 gives rise to pairs of magnetic energy
levels that often have markedly different relaxation properties.
Though there are early studies that have clarified some of the
issues regarding sodium relaxation, there is a need for NMRD studies
of sodium in tissue.

Fluorine, with a Larmor frequency close to that of protons,
is a possible exogenous nucleus for NMR imaging, particularly of
the circulatory system because of the steady development of blood
substitutes containing fluorinated hydrocarbons. Though the fluorine
nucleus is magnetically much like the proton, fluorine can be anionic
and thus can bind readily in the first coordination sphere of (paramagnetic) transition metal ions, with the attendant possibility of a
large contact contribution to its relaxivity. It is known, for
example, that the relaxation rates of fluorine nuclei are exquisitely
sensitive to the presence of Cu^{2+} ions [54,55]; indeed, so much so
that a shortened relaxation time can often be diagnostic for certain
copper-containing proteins [55]. Few NMRD data for fluorine have
been reported to date. However, they should be particularly interesting, and complex at low fields, given the possibility of extensive
cross-relaxation with protons due to overlapping of their respective
magnetic energy levels.

5.3. Instrumentation

No reference has been made in the preceding to any details of the field-cycling instrumentation, either its design or its capabilities. Although this is not the place for an in-depth description of the technology (indeed, a brief description of the instrumentation has been given [56] and the principles of field cycling have recently been reviewed [57]), certain details are germane to understanding the extent to which studies of relaxometry in tissue can be extended in new directions.

Relaxation rates can be measured with high accuracy and sensitivity at low fields by the field-cycling method introduced by Redfield [58,59]. In this method, the sample is allowed to reach its equilibrium magnetization in a high magnetizing field, 50 MHz at present. The field, produced by a highly regulated current in a liquid nitrogen-cooled air core (solenoid), is then reduced rapidly to the "measure field" at which T_1 is to be determined and maintained there for a "measure time" that represents a single point along the time-dependent magnetization decay; the field is then rapidly switched to the resonance field (25 MHz in the latest system), and the magnetization is measured using a 90-180° spin-echo sequence. This entire cycle is repeated for (typically 15) different values of the measure time between 0 and 1.5 T_1, to follow the magnetization decay, and again repeated several times for measure times of the order of 5 T_1 to obtain the equilibrium magnetization in the measure field. T_1 for the measure field is then computed by a weighted least squares fit of this set of data to a single exponential. In this way, rates at measure fields as low as 0.01 MHz can be obtained with sensitivities determined by the relatively high magnetizing and resonance fields. Values for T_1 at measure fields below 25 MHz are measured as described above. For higher measure fields, the field is set to zero during the first (magnetizing) part of the cycle, and the growth of the magnetization during the subsequent measuring period is monitored. Regardless of the value of the measure field, the resonance

field remains fixed (at 25 MHz) and no retuning of the radio frequency system is necessary. This procedure is readily automated to
obtained $1/T_1$ NMRD profiles at any sequence of field values from
0.01 to 60 MHz.

The instrumentation was developed to measure water proton
relaxation in biochemical solutions so that maximum sensitivity was
not stressed; samples are typically 0.6 ml, but samples in which
solvent protons have been replaced by 98% D_2O have been measured,
as have needle biopsy tissue samples, with reduced accuracy in both
instances. The emphasis has been on stability, reproducibility, and
automation, necessitating good temperature regulation of the sample,
computer control of most aspects of the measurement protocol, and a
highly linear signal detector, operating in the noncoherent mode
dictated by the field-cycling approach.

At present, a freon bath allows the variation of sample temperature from -10 to 40°C, with regulation to ±0.1°C. The field-
switching and settling time is relatively rapid, about 1 MHz/msec,
allowing rates as high as 35 $\sec^{-1}$ to be measured reliably, typically
better than ±1%, and higher rates to be measured with reduced accuracy. A typical NMRD profile, with 15 field values, requires about
20 min of unattended running time. No advance knowledge of the rates
is required, but an estimate in advance speeds data collection. The
field range is 0.01-60 MHz, limited at the high end by power dissipa-
tion in the magnet and at the low end by overshoot when approaching
the measure field, which can introduce nonadiabatic effects in the
magnetization amplitude.

By judicious redesign, the extremes of the field range could
be extended by about an order of magnitude at the low end and perhaps
to 80 MHz at the high end, with some loss in the versatility and ease
of operation of the relaxometer. Other specialized changes are also
possible: a larger working volume to permit the study of perfused
tissues, with the sacrifice of field range and temperature stability;
other pulse sequences to measure $1/T_2$ NMRD profiles and diffusion
constants; and more complex programming to permit routine collection
and examination of the magnetization data for multiexponential
behavior.

When functioning properly, the rate-limiting step in any experiment rapidly becomes dependent on the ease with which the accumulated data can be manipulated, filed, recalled, reduced, fit, and displayed. Accordingly, a significant amount of time has been devoted to the software needed for this aspect, which is now an integral, if not totally integrated, part of the system.

6. CONCLUDING REMARKS

The current expanding impact of NMR as an imaging modality in clinical medicine and its possible extension to NMR microscopy in, for example, histology and pathology, rests on the fact that relaxation rates of tissue protons depend on tissue type, that these rates can be manipulated by the presence of either endogenous or exogenous contrast-enhancing agents, and, hopefully, that relaxation rates of tissue protons and their response to such agents will be altered by the presence of disease.

Our experience to date has shown that what has been learned from the relaxometry of homogeneous in vitro solutions of proteins, paramagnetic ions, and their many complexes is directly applicable to the study of excised tissue, which in turn behaves in most instances as it does in vivo, as observed by NMR imaging. However, when such studies of tissue are made in vitro, the field and temperature dependencies of the relaxation rates can be measured, and in many cases the chemical state of the paramagnetic agent can be ascertained. Hence the importance and the relevance to biological systems of studies of the relaxometry of paramagnetic ions in tissue.

ABBREVIATIONS

DOTA   1,4,7,10-tetraazacyclododecane-N,N',N'',N'''-tetraacetic acid
DTPA   diethylenetriamine pentaacetate
EDTA   ethylenediaminetetraacetate
EHPG   ethylenebis(2-hydroxyphenylglycinate)

NOTA   1,4,7-triazacyclononane-N,N',N''-triacetic acid

PDTA   propanoldiaminetetraacetate

REFERENCES

1. S. H. Koenig and R. D. Brown III, *Magn. Reson. Med., 1,* 437 (1984).

2. S. H. Koenig, R. D. Brown III, E. J. Goldstein, K. R. Burnett, and G. L. Wolf, *Magn. Reson. Med., 2,* 159 (1985).

3. M. Spiller, S. H. Koenig, G. L. Wolf, and R. D. Brown III, *Invest. Radiol.,* in press.

4. S. H. Koenig and R. D. Brown III, *Magn. Reson. Med., 1,* 478 (1984).

5. S. H. Koenig, C. Baglin, R. D. Brown III, and C. F. Brewer, *Magn. Reson. Med., 1,* 496 (1984).

6. S. H. Koenig and R. D. Brown III, *Invest. Radiol., 20,* 297 (1985).

7. S. H. Koenig, R. D. Brown III, D. Adams, D. Emerson, and C. G. Harrison, *Invest. Radiol., 19,* 76 (1984).

8. "Contrast Enhancement in Biomedical NMR: A Symposium" (G. L. Wolf, ed.), *Physiol. Chem. and Phys., 16* (1984).

9. R. B. Lauffer, T. J. Brady, R. D. Brown III, C. Baglin, and S. H. Koenig, *Magn. Reson. Med., 3* (1986), in press.

10. H. F. Bennett, R. D. Brown III, S. H. Koenig, and H. M. Swartz, *Magn. Reson. Med.,* submitted.

11. P. A. Bottomley, T. H. Foster, R. E. Argersinger, and L. M. Pfeifer, *Med. Phys., 11,* 425 (1984).

12. T. R. Lindstrom and S. H. Koenig, *J. Magn. Reson., 15,* 344 (1974).

13. T. R. Lindstrom, S. H. Koenig, T. Boussios, and J. F. Bertles, *Biophys. J., 16,* 679 (1976).

14. S. H. Koenig, R. D. Brown III, and T. R. Lindstrom, *Biophys. J., 34,* 397 (1981).

15. S. H. Koenig, C. M. Baglin, and R. D. Brown III, *Magn. Reson. Med., 2,* 283 (1985).

16. R. D. Dowsing and J. F. Gibson, *J. Chem. Phys., 50,* 294 (1969).

17. R. Aasa, *J. Chem. Phys., 50,* 294 (1969).

18. R. C. Brasch, G. E. Wesbey, C. A. Gooding, and M. A. Koerper, *Radiology, 150,* 767 (1984).

19. R. B. Lauffer, W. L. Grief, D. D. Stark, A. C. Vincent, S. Saini, V. J. Wedeen, and T. J. Brady, *J. Comput. Assist. Tomogr., 9,* 431 (1985).

20. G. E. Wesbey, R. C. Brasch, B. L. Engelstad, A. A. Moss, L. E. Crooks, and A. C. Brito, *Radiology, 149,* 175 (1983).

21. J. Bloch and G. Navon, *J. Inorg. Nucl. Chem., 42,* 693 (1980).

22. H. Pfeifer, *Ann. Phys. (Leipzig), 8,* 1 (1961).

23. H. Pfeifer, *A. Naturforsch. A, 17,* 279 (1962).

24. J. H. Freed, *J. Chem. Phys., 68,* 4034 (1978).

25. J. P. Albrand, M. C. Taieb, P. H. Fries, and E. Belorizky, *J. Chem. Phys., 78,* 5809 (1983).

26. P. Aisen and I. Listowsky, *Ann. Rev. Biochem., 49,* 357 (1980).

27. S. H. Koenig and W. E. Schillinger, *J. Biol. Chem., 244,* 6520 (1969).

28. S. H. Koenig and R. D. Brown III, in *The Coordination Chemistry of Metalloenzymes* (I. Bertini, R. S. Drago, and C. Luchinat, eds.), Reidel, Dordrecht, 1983, p. 19.

29. I. Bertini, F. Briganti, S. H. Koenig, and C. Luchinat, *Biochemistry, 24,* 6287 (1985).

30. S. H. Koenig, R. D. Brown III, I. Bertini, and C. Luchinat, *Biophys. J., 41,* 179 (1983).

31. C. A. Clegg, J. E. Fitton, P. M. Harrison, and A. Treffry, *Prog. Biophys. Molec. Biol., 36,* 56 (1980).

32. L. P. Rosenberg and N. D. Chasteen, in *The Biochemistry and Physiology of Iron* (P. Saltman and J. Henenaur, eds.), Elsevier Biomedical, New York, 1982, p. 405ff.

33. S. H. Koenig, R. D. Brown III, J. F. Gibson, T. J. Peters, and R. A. Ward, *Magn. Reson. Med., 3* (1986), in press.

34. M. Gueron, *J. Magn. Reson., 19,* 58 (1975).

35. A. J. Vega and D. Fiat, *Mol. Phys., 31,* 347 (1976).

36. H. Levanon, G. Stein, and Z. Luz, *J. Chem. Phys., 53,* 876 (1970).

37. G. Laukien and J. Schlüter, *Z. Physik, 146,* 113 (1956).

38. N. Bloembergen, *J. Chem. Phys., 27,* 572 (1957).

39. M. Cohn, *Quart. Rev. Biophys., 3,* 61 (1970).

40. A. R. Peacocke, R. E. Richards, and B. Sheard, *Mol. Phys., 16,* 177 (1969).

41. A. Danchin and M. Gueron, *J. Chem. Phys., 53,* 3599 (1970).

42. J. Eisinger, R. G. Shulman, and B. M. Szymanski, *J. Chem. Phys., 36,* 1721 (1962).

43. S. H. Koenig, R. D. Brown III, and J. Studebaker, *Cold Spring Harbor Symp. Quant. Biol.*, *36*, 551 (1971).

44. C. F. Brewer, R. D. Brown III, and S. H. Koenig, *J. Biomol. Struct. Dynam.*, *1*, 961 (1983).

45. J. Oakes and E. G. Smith, *J. Chem. Soc. Faraday Trans. 2(77)*, 299 (1981).

46. C. F. G. C. Geraldes, A. D. Sherry, R. D. Brown III, and S. H. Koenig, *Mag. Reson. Med.*, *3*, 242 (1983).

47. M. Spiller, unpublished.

48. R. Kurland and S. H. Koenig, unpublished.

49. S. H. Koenig and M. Epstein, *J. Chem. Phys.*, *63*, 2279 (1975).

50. S. H. Koenig and R. D. Brown III, *J. Magn. Reson.*, *61*, 426 (1985).

51. J. Reuben, *Biochemistry*, *10*, 2834 (1971).

52. P. O'Hara and S. H. Koenig, *Biochemistry*, *25*, 1445 (1986).

53. J. F. Desreux and S. H. Koenig, unpublished.

54. M. Eisenstadt and H. L. Friedman, *J. Chem. Phys.*, *48*, 4458 (1968).

55. A. Rigo, P. Viglino, E. Argese, M. Terenzi, and G. Rotilio, *J. Biol. Chem.*, *254*, 1759 (1979).

56. S. H. Koenig and R. D. Brown III, in *NMR Spectroscopy of Cells and Organisms* (R. K. Gupta, ed.), CRC Press, Boca Raton, FL, 1986, in press.

57. F. Noack, *Prog. NMR Spectr.*, *18*, 171 (1986).

58. A. G. Anderson and A. G. Redfield, *Phys. Rev.*, *116*, 583 (1959).

59. A. G. Redfield, W. Fite, and H. E. Bleich, *Rev. Sci. Instrum.*, *39*, 710 (1968).

Author Index

Numbers in parentheses are reference numbers and indicate that an author's work is referred to although his name may not be cited in the text. Underlined numbers give the page on which the complete reference is listed.

A

Abragam, A., 4(14), 6(25), 17 (14), <u>41</u>, <u>42</u>; 69(63), <u>85</u>
Abraham, R. J., 193(27,28), 194 (27-29), 208(27,28), <u>224</u>
Adams, D., 232(7), 261(7), <u>268</u>
Aisen, P., 99(30), 100(30), 102 (30), 109(59,62,64), 110(30), 111(30,66), 112(59,69), <u>117</u>, <u>119</u>; 238(26), <u>269</u>
Albrand, J. P., 238(25), <u>269</u>
Alden, R. A., 123(109), 171 (108,109), <u>184</u>
Alsaadi, B. M., 71(76), <u>85</u>
Al'tshuler, S. A., 69(62), <u>85</u>
Amzel, L. H., 34(101), <u>45</u>
Anderson, A. G., 265(58), <u>270</u>
Anderson, R. R., 155(78), <u>182</u>
Anderson, T., 197(39), <u>225</u>
Andres, P. J., 16(49), <u>43</u>
Anglister, J., 34(109-111), 35 (109), 36(109,110), 37(110), 38(110), 39(110), 40(110), <u>45</u>
Angström, J., 173(125), <u>185</u>
Antanaitis, B. C., 99(30), 100 (30), 102(30), 109(59,62,64), 110(30), 111(30,66), 112(59, 69), <u>117</u>, <u>119</u>
Antonini, E., 127(23), <u>180</u>; 197(38), <u>225</u>
Argersinger, R. E., 232(11), <u>268</u>
Argese, E., 264(55), <u>270</u>
Argos, P., 123(25), 131(25), <u>180</u>; 215(72), <u>226</u>
Armstrong, R. N., 23(68), <u>44</u>

Armstrong, W. H., 98(38), 100(37, 38), 104(37,38), <u>117</u>
Asasa, R., 233(17), 240(17), <u>268</u>

B

Baglin, C., 66(58), <u>85</u>; 232(5,9), 233(15), 234(15), 236(15), 243 (15), 246(5), 251(5), <u>268</u>
Bailey, D. B., 96(17), <u>116</u>
Balakrishnan, K., 34(108), <u>45</u>
Balch, A. L., 100(34), <u>117</u>; 141 (46), <u>181</u>
Baldwin, J. E., 211(66,67), 212 (68), <u>226</u>
Banci, L., 52(9), 61(37,38), 62 (37), 64(38), 65(38,57), 67(38), 75(81), 76(81), 77(81), 78(81), <u>83</u>, <u>84</u>, <u>85</u>, <u>86</u>
Baram, A., 7(28), <u>42</u>; 58(16), 71 (16), 72(16), <u>83</u>
Barfield, M., 15(46), <u>43</u>
Barrett, J., 126(12), <u>179</u>
Barry, C. D., 190(17), 199(17), 206(17,59,60), 207(17), <u>224</u>, <u>226</u>
Bartsch, R. G., 14(41), <u>42</u>; 68 (59), <u>85</u>; 137(41), <u>181</u>
Bates, R. D., Jr., 29(87), 30(87), <u>44</u>
Baumbach, G. A., 109(60), <u>119</u>
Bazer, F. W., 109(60,64), 111(65), <u>119</u>
Bearden, A. J., 97(23), 103(45), 114(45), <u>117</u>, <u>118</u>

Subject Index